A FUNCTIONAL BIOLOGY OF STICKLEBACKS

FUNCTIONAL BIOLOGY SERIES

General Editor: Peter Calow, Department of Zoology,
University of Sheffield

A Functional Biology of Free-living Protozoa
Johanna Laybourn-Parry

A Functional Biology of Sticklebacks

R.J. WOOTTON

Department of Zoology,
The University College of Wales, Aberystwyth

UNIVERSITY OF CALIFORNIA PRESS
Berkeley and Los Angeles

© 1984 R.J. Wootton

University of California Press
Berkeley and Los Angeles, California

Library of Congress Cataloging in Publication Data

Wootton, R.J. (Robin Jeremy)
   A functional biology of sticklebacks.

   Bibliography: p.
   Includes index.
   1. Sticklebacks. I. Title.
   QL638.G27W663   1984   597'.53   84-8864
   ISBN 0-520-05381-8

Printed and bound in Great Britain

CONTENTS

For Siobhan and Sean, sceptical of sticklebacks

FUNCTIONAL BIOLOGY SERIES: FOREWORD

General Editor: Peter Calow, Department of Zoology,
University of Sheffield, England

The main aim of this series will be to illustrate and to explain the way organisms 'make a living' in nature. At the heart of this — their *functional biology* — is the way organisms acquire and then make use of resources in metabolism, movement, growth, reproduction, and so on. These processes will form the fundamental framework of all the books in the series. Each book will concentrate on a particular taxon (species, family, class or even phylum) and will bring together information on the form, physiology, ecology and evolutionary biology of the group. The aim will be not only to describe *how* organisms work, but also to consider *why* they have come to work in that way. By concentrating on taxa which are well known, it is hoped that the series will not only illustrate the success of selection, but also show the constraints imposed upon it by the physiological, morphological and developmental limitations of the groups.

Another important feature of the series will be its *organismic orientation*. Each book will emphasise the importance of functional *integration* in the day-to-day lives and the evolution of organisms. This is crucial since, though it may be true that organisms can be considered as collections of gene-determined traits, they nevertheless interact with their environment as integrated wholes and it is in this context that individual traits have been subjected to natural selection and have evolved.

The key features of the series are, therefore:

(1) Its emphasis on whole organisms as integrated, resource-using systems.

(2) Its interest in the way selection and constraints have moulded the evolution of adaptations in particular taxonomic groups.

(3) Its bringing together of physiological, morphological, ecological and evolutionary information.

P. Calow

PREFACE

Writing this book in the *Functional Biology Series* has allowed me to combine two of my major academic interests, research on the biology of the sticklebacks and teaching courses on theoretical ecology. The purposes of the book are twofold. The first is to demonstrate that the theoretical framework in ecology and evolutionary biology that has been developed, much of it over the past two decades, can be used to illuminate our understanding of the ways in which animals function in everyday life. The second is to show that the knowledge that can only be gained by a close and detailed study of a taxon will be required to test critically this theoretical framework.

In writing this volume, I have been greatly helped by the advice of my colleagues. I would like to thank P. Calow, G.J. FitzGerald and F.A. Huntingford, who commented on the complete manuscript. M.A. Bell, J. Gee, H. Guderley, M. Milinski, D. Wharton and F. Whoriskey read sections of it. Their criticisms improved both the accuracy and clarity of the text. The mistakes of fact and interpretation that remain are my responsibility. I also thank Denise Long for her excellent drawings and R. Matthews for help with the word-processing. Finally, I thank my wife, Maureen, whose help and advice at all stages in the writing of this book were invaluable.

R.J. Wootton
Aberystwyth

ACKNOWLEDGEMENTS

All the Figures were freshly drawn, but I am grateful to the following for their permission to adapt copyright material. Figure 1.1 from *Amer. Nat.* **16**, 775-84 (1976), Dr E.R. Pianka and the American Society of Zoologists. Figures 2.7 and 7.5 with permission from *The Biology of the Sticklebacks*; copyright: Academic Press, London. Figure 4.1 from *Neth. J. Zool.* **28**. 485-523 (1978), Dr G.Ch. Anker and E.J. Brill, Leiden. Figure 4.2 from *Behaviour* **15**, 284-318 (1960), Dr B. Tugendhat and E.J. Brill, Leiden. Figure 4.3 from *Z. Tierpsychol.* **45**, 373-88 (1977), Dr M. Milinski and Paul Parey (Berlin). Figure 4.4 reprinted by permission from *Nature* **275**, 642-44, Copyright © The Macmillan Press Ltd 1978 and Dr M. Milinski. Figure 4.5 from *J. exp. Mar. Biol. Ecol.* **25**, 151-58 (1976), Dr R.N. Gibson and Elsevier Biomedical Press. Figures 6.2–6.4 from *J. Fish Biol.* **20**, 409-22 (1982), Fisheries Society of the British Isles. Figure 7.3 R.A. Moore. Figure 8.1 from *Behaviour* **10**, 205-37 (1957), E.J. Brill, Leiden. Figure 10.3 from *Amer. Nat.* **117**, 113-32 (1981), Dr M.A. Bell and © 1981 The University of Chicago. Figure 10.5 from *Evolution* **33**, 641-48 (1979), Dr D.W. Hagen and the Editor, *Evolution*. Figure 10.6 from *J. Fish. Res. Bd Canada* **26**, 3183-208, Fisheries and Oceans, Canada.

1 INTRODUCTION

The past two decades have seen a proliferation of theoretical studies in the related fields of ecology and evolutionary biology. Studies of optimal foraging, demography and life-history strategies spring quickly to mind. A strengthening of the theoretical framework of these fields is welcome, but it does bring with it an attendant danger. Mathematical equations can have a seductive elegance, simulation models developed on a computer a persuasive productivity. The danger develops when these theoretical developments are not tested by encounters with the often recalcitrant living organisms whose biology the theories should illuminate. The theoretical developments lose touch with the biological reality (Stearns, 1976, 1977).

Even when theories are tested, the diversity of living organisms is such that it may be possible to select a suitable example which seems to confirm the favoured theory. A better test of the adequacy of the theoretical framework is to apply it to organisms whose biology is already well known. How well can the framework be used to interpret and understand the life histories of active, living, evolving organisms coping with a difficult and changing world? Such an approach is also important in showing students that the theoretical framework can profitably be used to understand the biology of real organisms. Often students find it difficult to relate the theory to what they see living organisms doing.

An animal can be viewed as a system that converts the energy and materials in its food into offspring. An input, food, is 'mapped' into an output, the progeny (Figure 1.1) (Pianka, 1976). The success of the animal is measured by the number of its offspring that survive to reach their sexual maturity. This success has to be achieved in the face of possible shortages of food, living space and time in a potentially hostile physical and biotic environment. Although the description of the life histories of animals has always formed a significant part of zoology, the quantitative analysis of life histories is a recent development. This analysis has two components: theoretical studies of the consequences and the adaptive significance of patterns of survivorship

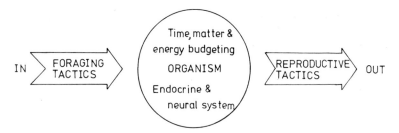

Figure 1.1: Organism as an Input-Output System 'Mapping' Food into Progeny. Modified from Pianka (1976)

and reproduction and empirical studies which attempt to identify the causal factors responsible for the patterns observed in nature. Two sets of causal factors are sought, those responsible for the evolution of the observed life history, the ultimate factors, and those responsible for the day-to-day control of the components of the life history, the proximate factors.

Unfortunately few species are studied in detail at each stage of their life history. Zoologists frequently study species which clearly illustrate some problem of immediate interest, while the other aspects of the biology of the species fade into the background. The significance of the features under study for the pattern of survivorship and reproduction of the species may even be ignored. A satisfactory analysis of the life history of a taxon also demands that the effects of experimental manipulations of environmental factors can be observed on the life history characteristics. Such manipulations are difficult or impossible for many species.

A species that can be studied at each stage of its life history, both in the field and under experimental conditions, is the common, euryhaline, teleost fish, the threespine stickleback, *Gasterosteus aculeatus*. This species came into prominence because of a classic series of ethological studies on the reproductive behaviour of the male (Tinbergen, 1951), but other aspects of its biology command attention. It is geographically widespread and is variable physiologically, morphologically and in its life-history characteristics. This variability makes the stickleback an ideal subject for the analysis of life-history patterns. The other species in the stickleback family (Gasterosteidae) (Table 1.1) have been less studied, but they provide useful comparisons which help clarify the probable adaptive significance of some of the life-history characteristics of *G. aculeatus*.

Although the biology of this family has been extensively reviewed

Table 1.1: Species in the Stickleback Family (Gasterosteidae, Teleostei)

Apeltes quadracus	The fourspine stickleback
Culaea inconstans	The brook stickleback
Gasterosteus aculeatus	The threespine stickleback
Gasterosteus wheatlandi	The blackspotted stickleback
Pungitius pungitius	The ninespine stickleback
Pungitius platygaster	The Ukrainian stickleback or small southern stickleback
Spinachia spinachia	The fifteenspine stickleback or the sea stickleback

(Wootton, 1976), no attempt was then made to interpret this biology in the context of a theoretical framework. The purpose of this book is to use the wealth of information that has been collected on the sticklebacks to illustrate how a picture of these fish, as they function in the real world, can be created by integrating both empirical data and the theoretical framework now available. Hopefully, such an approach will be applied to other well-studied groups of animals and plants, so that the symbiotic relationship between theoretical advances and empirical studies will be strengthened in ecological and evolutionary biology.

Using the input-output model shown in Figure 1.1 as its major structural theme, this book describes how sticklebacks use their spatial habitat, food supply and time during their life and the patterns of growth, survivorship and reproduction that result. The successful 'mapping' of resources into progeny is mediated by the time and energy budgeting of the stickleback, budgeting which is controlled by physiological mechanisms within the constraints set by morphological design. The success of this mapping can be judged by the abundance and geographical distribution of the sticklebacks. Interactions of sticklebacks with other species, especially predators, parasites and competitors, influence the patterns of growth, survivorship and reproduction. But such interactions can also lead to evolutionary changes and the sticklebacks prove to be unusually suitable material for an analysis of the action of natural selection and other evolutionary processes on a vertebrate taxon. In a final chapter, the life history patterns of the sticklebacks are discussed in the context of theoretical studies on the evolution of life-history strategies.

2 SPATIAL DISTRIBUTION

Introduction

Spatial distribution of the sticklebacks has several components: the global distribution of the family, the global distributions of the individual genera and species, and the local distributions of individual populations. A related factor is the extent to which the spatial distribution of populations reflects migrations between different habitats and localities. At one extreme, the global patterns reflect the evolutionary and adaptational history of the family and its species, whereas, at the other end of the scale, local distributions depend on the behavioural choices made by individual fish during their ontogeny. These behavioural choices are constrained by the physiological capacity of the fish to adapt to new environmental factors (Chapter 5).

Global Distribution of the Family

The geographical range of the family is immense. Representatives are found throughout much of the northern hemisphere above about 35°N, with the exception of the continental heartland of Asia. In North America, Gasterosteids occur from the Great Lakes northwards, extending further south on both the eastern and western seaboards. Two species also occur in Greenland. The family is found in most of western Europe, and as far east as the Caspian and Aral Seas and as far north as the northern coastline of Siberia. In Asia, their distribution is restricted to coastal areas and islands southwards to Korea. Over large areas of this distribution two or more species are sympatric. This sympatry reaches a peak in north-eastern America, where, in the area of the St. Lawrence Estuary, the ranges of five species overlap. North America also holds the highest number of genera: *Gasterosteus*, *Pungitius*, *Culaea* and *Apeltes*. Three genera are found in Europe: *Gasterosteus*, *Pungitius* and *Spinachia*. Whereas in Asia only two genera occur, *Gasterosteus* and *Pungitius*. Because the evolutionary relationships

4

between these genera are not understood, the events which have yielded this global pattern are not known. The teleost family to which the Gasterosteids are most closely related, the tube-snouts (Aulorhynchidae), is found only in the Pacific Ocean (Wootton, 1976).

Global Distribution of the Genera

Gasterosteus. The best known of the sticklebacks, *G. aculeatus*, is found on all three northern continents, and on all three its distribution is essentially coastal (Figure 2.1). In western North America, it is found along the Pacific coastline from Alaska to southern California. In the east it runs from Hudson's Bay and Baffin Island to Chesapeake Bay, and penetrates down the St. Lawrence Estuary to Lake Ontario. On the Pacific coast, the southern populations are confined to fresh water, but on the Atlantic coast, freshwater populations are rare south of Maine. It is not found along the northern coastline of North America, nor does it penetrate into the interior of the continent. In Asia, it is also absent from the northern coastline, but occurs in the Bering Straits and reaches as far south as Japan and Korea. Again there is no penetration into the continental heartland. Its distribution in Europe includes Iceland and as far north as the White Sea. In the south it occurs in Spain, Sardinia, most of Italy and parts of the Balkans peninsula. It is also found along the coast of the Black Sea and northwards into Poland and western Russia. It is not found east of the Black Sea nor in the Volga River. Its deepest penetration into Europe probably occurs down the Rhine, along the Dneister and Dneiper in south-east Europe, and along the Elbe, Oder and Neisse in north-eastern Europe. In Greenland, it is found on both the eastern and western coastlines up to about 70°N. Throughout this global distribution, there is a complex pattern of polymorphisms, which will be considered in Chapter 10.

A second species of *Gasterosteus*, *G. wheatlandi*, is endemic to a relatively small area of the Atlantic seaboard of North America, including Newfoundland, the mouth of the St. Lawrence Estuary and as far south as New Jersey. Throughout this range it is sympatric with *G. aculeatus* (Figure 2.2).

Pungitius. This genus is also found on all northern continents (Figure 2.3), and although sympatric with *Gasterosteus* over large areas, has a distinctive global distribution, whose 'centre of gravity' is more

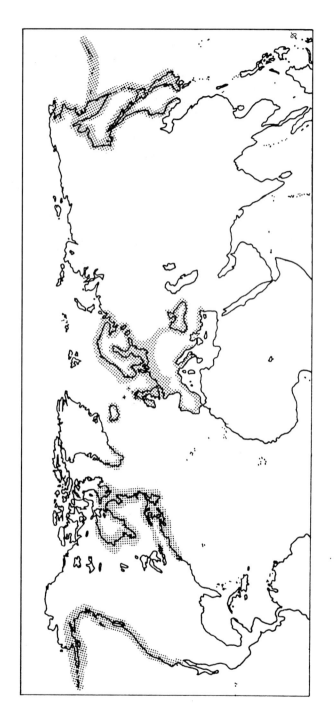

Figure 2.1: Approximate Global Distribution of *Gasterosteus aculeatus*

Figure 2.2: Approximate Distribution of *Gasterosteus wheatlandi*

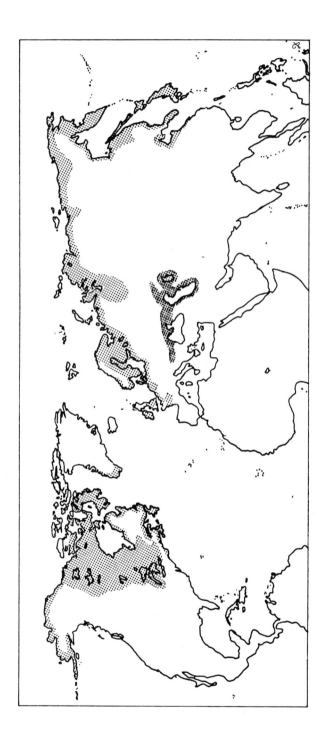

Figure 2.3: Approximate Global Distribution of *Pungitius*. Light Stippling, *P. pungitius*; Heavy Stippling, *P. platygaster*

northerly. In North America *P. pungitius* is found in coastal regions of Alaska and along the northern coastline of the continent, on Baffin Island, then around Hudson's Bay and the coast of Labrador, reaching as far south as New Jersey. It is also found deep in the continent, with a distribution stretching north-west from the Great Lakes. Throughout much of this continental distribution it is sympatric with *Culaea*. It is not found on the Pacific coast south of Alaska. In Asia its distribution is entirely coastal. Three distinct sub-species are recognised within this Asian distribution: *P.p.pungitius* occurs on the northern coastline of Siberia and the Kamchatka Peninsula and as far south as Japan; *P.p.sinensis* ranges from the southern coastline of Kamchatka to the Yangtse River and includes Japan in its range; *P.p.tymensis* is restricted to the islands of Sakhalin and Hokkaido and the west coast of the Sea of Japan. Over much of this range, *Pungitius* is sympatric with *Gasterosteus*. There is a continuous distribution of *P. pungitius* along the northern coastline of the Eurasian continental mass, which contrasts with the absence of *G. aculeatus* along this coastline. In Europe, *P. pungitius* ranges from the Arctic Ocean to the River Loire in France. The distribution is essentially coastal, so in Eurasia *P. pungitius* does not penetrate into the centre of the landmass as it does in North America. Throughout Europe there is sympatry with *G. aculeatus*. A second species of *Pungitius*, *P. platygaster*, is found in the river systems that drain into the Black, Caspian and Aral Seas.

Culaea. This is one of the two genera restricted to North America (Figure 2.4). *Culaea inconstans*, in contrast to all the other sticklebacks, has a distribution that is largely continental, not coastal. It is found approximately between latitudes 40 and 60°N, east of the Rocky Mountains, from the Great Lakes north to Great Slave Lake. It reaches the Atlantic seaboard in Hudson's Bay and in the St. Lawrence Estuary. Throughout almost all of this range, *C. inconstans* is sympatric with *P. pungitius*, and in the east with *G. aculeatus*.

Apeltes. As shown in Figure 2.5, *A. quadracus* is restricted to the Atlantic coast of North America, from the Gaspé Basin in Quebec in the north to Virginia in the south. Over much of this range *A. quadracus* is sympatric with *G. aculeatus*, *G. wheatlandi* and *P. pungitius*.

Spinachia. The largest and most elegant of the sticklebacks, *S. spinachia* is entirely restricted to the coastal waters of western Europe, from the Bay of Biscay in the south to northern Norway (Figure 2.6). It does not penetrate into fresh water anywhere within this range.

Figure 2.4: Approximate Distribution of *Culaea inconstans*

Figure 2.5: Approximate Distribution of *Apeltes quadracus*

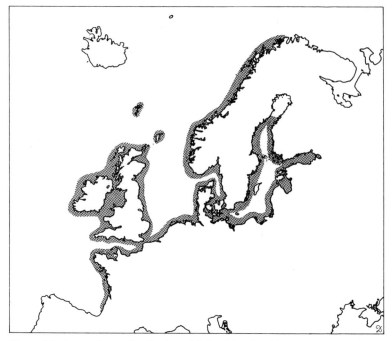

Figure 2.6: Approximate Distribution of *Spinachia spinachia*

The Habitats of the Sticklebacks

Although the degree of geographical sympatry between the members of the stickleback family is extensive, the spatial overlap between local populations of the different species will depend on the habitat preference of the species within a geographical area.

Habitat Preferences of Gasterosteus

Gasterosteus aculeatus populations show one of two types of spatial behaviour. Some populations are resident in fresh water throughout the year, whereas other populations are anadromous, migrating to the sea in autumn and then migrating back into rivers, salt marshes or tidal pools in the spring to breed. River systems may contain both resident and anadromous populations. The leisurely mode of swimming of the stickleback excludes it from fast-flowing waters so the species is typically found in the still or slow-flowing water of lakes, ponds, lowland streams and sheltered coastal bays. Within this constraint, it lives in a wide variety of habitats ranging from large freshwater lakes to small ponds and streams. In some lakes, its spatial distribution is largely littoral but in others it is limnetic. Indeed, in some lakes, both littoral and limnetic populations may coexist (Chapter 10). The anadromous populations that migrate to the sea usually stay inshore, but sticklebacks have been found at distances up to 500km from the shore (Jones and John, 1978). Cod caught at depths of 200m and 160km from the coast of northern Europe had sticklebacks in their stomachs (Brown and Cheng, 1946). In the Barents Sea, sticklebacks can be abundant in the pelagic zone, usually associated with patches of floating algae (Parin, 1968).

Within its limited distribution, *G. wheatlandi* is found predominantly in coastal waters, although it does occur at a few freshwater sites. During late spring and summer in the Rivière des Vases, a small river in the St. Lawrence Estuary, *G. wheatlandi* was only captured near the mouth of the river, where the salinity varied between 3 and 20ppt. None was found upstream in fresh water, nor was any collected in freshwater pools in the locality, although they were in brackish-water pools. At the same locality, *G. aculeatus* and *P. pungitius* were found in the river mouth, in the fresh water upstream and in freshwater pools (Worgan and FitzGerald, 1981a).

Habitat of Pungitius

Pungitius also has anadromous and resident populations. Its habitat preference is similar to that of *Gasterosteus* — the two can often be collected in the same net — but *Pungitius* probably prefers areas of denser vegetation than *Gasterosteus* (Wheeler, 1969). Collections of both species in eastern England suggested that *P. pungitius* was better adapted to shallow, weedy, eutrophic waters that became depleted of oxygen, whereas *G. aculeatus* preferred more open, well oxygenated waters (Lewis *et al.*, 1972). In Rivière des Vases, the distribution of *P. pungitius* both in the river and in the salt-marsh pools was similar to that of *G. aculeatus* and included both the freshwater reaches of the river and freshwater pools. Adults of both species of *Gasterosteus* and *P. pungitius* had left the pools by early July, although the latter species persisted in the river until the end of August before dispersing into the St. Lawrence Estuary (Worgan and FitzGerald, 1981a). In Lake Huron, *P. pungitius* occurred close to the bottom at depths between 10 and 15m during the day, but showed a tendency to move deeper at night (Emery, 1973). Breeding *P. pungitius* in Lake Huron were observed among rocks around the shore (McKenzie and Keenleyside, 1970), but in Lake Superior, sexually mature fish were collected over highly organic muds at depths between 20 and 45 m (Griswold and Smith, 1973).

Habitat of Culaea

Sympatric with freshwater populations of *P. pungitius* over a wide geographical area, *C. inconstans* also has similar habitat preferences. It lives in heavily weeded, cool waters, including streams and ponds, as well as the swampy margins and bogs of lakes. It is found in prairie lakes which may become relatively deoxygenated in winter when covered with ice and snow, but has behavioural and physiological adaptations to survive these conditions (Chapter 5).

Habitat of Apeltes

Most populations of *A. quadracus* occur in vegetated areas with calm water. Populations commonly inhabit brackish estuaries and lagoons, whereas fully marine populations are rarer and tend to occur in sheltered inlets or mudflats where there is the eel-grass, *Zostera*. There are also resident freshwater populations throughout most of its geographical range, but it is probably not found in fresh waters of granitic areas where the water has a very low salt content (Blouw and Hagen, 1981). In Rivière des Vases, *A. quadracus* extended further up the river than *G. wheatlandi* but did not enter the freshwater stretches, nor was it found in any of the salt-marsh pools in which the other three stickleback species were breeding. It persisted in the river until the autumn (Worgan and FitzGerald, 1981a). In a small freshwater stream in Connecticut, *A. quadracus* was only collected in stretches containing the plant *Elodea* and where the current was slack. When fish from this stream were given a choice, they showed a strong preference for the half of a tank that contained *Elodea* rather than the half containing *Potamogeton*, another plant from the stream (Baker, 1971).

Habitat of Spinachia

Spinachia spinachia is a fish of shallow, marine waters. Beds of *Zostera*, weed-covered rocks, and stands of fucoid seaweeds are its preferred habitats. Although it lives in water with a salinity as low as 5 ppt in the Baltic Sea, it is the only stickleback species that does not penetrate into fresh water. Although geographically it is sympatric with *G. aculeatus* and *P. pungitius*, there is probably little or no spatial overlap during the breeding season.

Migration in the Sticklebacks

Resident freshwater populations of *Gasterosteus*, *Pungitius* and *Culaea* may show a migration from the deeper water in which they over-wintered into the shallower waters of small streams and backwaters where they will breed, but the most interesting migration is that shown by the anadromous stickleback populations. These migrate downstream to the sea soon after the completion of the breeding season in summer, and overwinter in the sea. In spring, the sexually maturing adults migrate back on to the breeding grounds, which may be in brackish or fresh water. This migration is a movement between waters of different salinities. In some species both resident and anadromous populations are found in some river systems, with the anadromous population typically moving into the lower reaches of the system during the spring to breed. There may be an area of the system in which both popu-lations are present, giving the potential for a gene-flow between the resident and anadromous populations (Chapter 10). The migration means that the population exploits one habitat during spring and summer while it is breeding and the young fish are growing rapidly, and another habitat during late autumn and winter. The possible advantages of this migratory behaviour will be considered later in the context of the whole life history of the species (Chapter 11), but the physiological mechanisms that permit this exploitation of two habitats differing in salinity must be considered.

The migration has two components: a change in behaviour so that the fish make directed movements either up or down a salinity gradient (discussed in Chapter 5) and a process of physiological adapta-tion to the changing salinity as the migration proceeds.

Osmoregulation in Sticklebacks

For sticklebacks, as for other teleosts in fresh water, the osmotic pressure and the concentrations of inorganic ions of the blood and other body fluids are higher than in the external medium. Conse-quently, the fish tends to gain water and to lose inorganic ions, so that to maintain a constant osmotic pressure and concentration of inorganic ions of its body fluids it must excrete the excess water and take up inorganic ions. In sea water the situation is reversed. Compared with sea water, the body fluids have a lower osmotic pressure and concentration of inorganic ions. The fish tends to lose water to its external environ-ment and to take up inorganic ions, so to maintain a constant internal environment the fish must acquire water and excrete the excess ions.

The main organs for osmo- and ionic-regulation are the kidneys, the gills and in some circumstances the intestine (Figure 2.7).

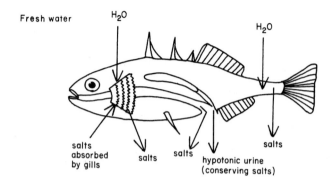

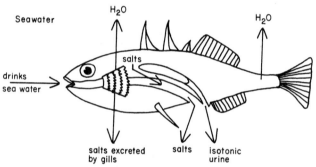

Figure 2.7: Direction of Flow of Water and Salts when a Euryhaline Fish is in Fresh Water and Sea Water. After Wootton (1976)

The trunk portion of the kidneys contains clusters of renal corpuscles, each of which consist of a Bowman's capsule with its associated glomerulus. Fluids filter from the tangle of blood capillaries that form the glomerulus into the lumen of the capsule, from where they pass into the kidney tubule via a neck region. From the neck region the filtrate passes down the first proximal and second proximal segments into the collecting tubule. All the collecting tubules from a cluster of corpuscles fuse and open into the ureter which leads to the urinary bladder. The epithelial cells of the proximal segments and probably those of the collecting ducts and ureter modify the composition of the filtrate by reabsorbing or secreting inorganic ions (Wootton, 1976). In fresh water, the urine produced by the kidneys is more dilute than the blood serum, indicating that ions have been reabsorbed from the filtrate during its passage to the bladder. In sea water, the flow rate of urine is reduced to conserve water and the urine is at the same concentration as the blood serum (Wendelaar Bonga, 1973).

The gills present a large surface area to the external environment which allows for the efficient exchange of oxygen and carbon dioxide, but they also present a large area through which water and inorganic ions can pass. In the epithelium of the gills are cells that are specialised for the transport of inorganic ions, the chloride cells. When the fish is in fresh water, the chloride cells probably take up inorganic ions, especially Na^+ and Cl^-, against the concentration gradient, but when the fish is in sea water, the cells excrete the excess inorganic ions again against the concentration gradient.

When the fish is in sea water, it drinks to replace the water it is losing. The water is absorbed through the walls of the alimentary canal, but some of the ions in the sea water are lost in the faeces and others are excreted via the chloride cells of the gills. An unusual situation arises in the sexually mature male adult, because the kidney of the male becomes modified to secrete the mucus that is used to glue the male's nest (Chapter 7). This transformation of the kidney probably reduces its effectiveness as an organ of osmoregulation, and in mature male *G. aculeatus* there is an increase in the rate of production of fluid from the intestine. This fluid is less concentrated than the blood serum and so provides one pathway by which the mature male can get rid of excess water while it is breeding in fresh water (De Ruiter, 1980).

Hormonal Control of Migration

The downstream migration usually takes place when the day length is shortening and the water temperatures are declining, whereas the upstream migration takes place as day lengths increase and water temperatures rise. The changing day lengths and temperatures provide cues which trigger changes in the central nervous system and endocrine organs to induce the migratory behaviour and the process of physiological adaptation to a changing salinity (Baggerman, 1957).

Some studies on *G. aculeatus* suggest that the thyroid gland which produces the hormone thyroxine is important in the control of migration. Anadromous *G. aculeatus* early in their upstream migration had highly active thyroids, and the lowest activity was shown by fish in the sea, although the activity of the gland declined during the breeding season and in the post-breeding phase while the fish were still in fresh water (Honma *et al.*, 1977). The production of thyroxine is controlled by a hormone released by the pituitary, and the thyrotropic cells in the pituitary were maximally developed in fish that were collected in the sea in spring just prior to their migration into fresh water (Leatherland, 1970). Baggerman (1957) found that anadromous *G. aculeatus* treated with thyroxine developed a preference for fresh water over brackish water, and the hormone may also help to regulate the metabolic changes associated with increased locomotion required during upstream migration (Honma *et al.*, 1977). When *G. aculeatus* in fresh water were fed with thyroid tissue, they showed disturbances in their ability to regulate the chloride content of the blood, and if treated in sea water many fish died (Koch and Heuts, 1942). As yet these studies do not provide a coherent picture of the specific role of thyroid hormones in the control of migration of the stickleback and, given the general effects that thyroxine has on metabolism as opposed to any specific effects it may have on migration, the teasing apart of general and specific effects will demand careful experimental studies.

A pituitary hormone that seems crucial in the ability of anadromous *G. aculeatus* to move into fresh water in spring is prolactin. Injections of prolactin significantly reduced the mortality in fish that were transferred from sea water into fresh water in early winter. After the transfer, injected fish were able to produce a dilute urine and better regulation of the concentrations of Na^+ and Cl^- in the blood, abilities which uninjected fish also showed during the spring at the time of their normal migration into fresh water (Lam, 1972). The cells in the pituitary that produce prolactin seem to be most active in spring and least active in winter in anadromous fish, but in *G. aculeatus* permanently resident in fresh water, these cells are active throughout the year (Leatherland, 1970; Benjamin, 1974). Prolactin accelerated the changes in the structure of the epithelium cells of the renal tubules of migratory *G. aculeatus* that took place when the fish were transferred from sea water to fresh water (Wendelaar Bonga, 1976). Prolactin also reduced the net loss of Na^+ and Cl^- from the gill region, and induced an increase in the density of mucus cells on the gills (Lam, 1972). An increase in the thickness of the epidermis and an increase of mucus cells in the skin have been correlated with the activity of the prolactin cells of the pituitary (Wendelaar Bonga, 1978a). This stimulation of the activity of prolactin cells when the stickleback is in fresh water seems to be induced by the low Ca^{2+} levels in the environment, rather than by low Na^+ or K^+ concentrations. The corpuscles of Stannius, small endocrine glands lying just behind and dorsal to the kidneys, probably play a role in the regulation of Ca^{2+} levels in the blood and so may also help to regulate the activity of the prolactin cells (Wendelaar Bonga, 1978a, b).

Dispersal of the Sticklebacks

With the ability of the stickleback to tolerate a wide range of salinities and the tendency towards migratory behaviour, the potential for dispersal along a coastline from river system to river system is high. The global distribution of both *G. aculeatus* and *P. pungitius* suggests that such dispersal has been extremely important, but the interpretation of the dispersal pattern in these two species is complicated by the marked polymorphism both in morphology and migratory behaviour that is shown by both species. This problem is taken up again in Chapter 10.

3 STRUCTURE AND FUNCTION

Introduction

Only a brief description of the morphology and anatomy of the stickle-backs will be given to provide the background for the other chapters. A more detailed description is given in Wootton (1976).

General Appearance

Four of the five genera are similar in appearance. They are small spindle-shaped fishes with a length-to-depth ratio that varies between about 4:1 and 5:1. They rarely exceed 100mm in length and are typically only half this. The exception is *Spinachia*, an elegant, elongated fish which can reach a length of 200mm, with a length-to-depth ratio of about 11:1. All sticklebacks have a relatively slender tail stalk (caudal peduncle) with a truncate tail fin. The dorsal and anal fins are set well back along the body, but in front of the dorsal fin is a series of dorsal spines whose number varies from species to species. The paired fins, the pectorals and pelvics, have evolved distinct functions. Each pectoral fin has a narrow base set on the side of the body, but the fins are broad and rounded and are used in the normal mode of locomotion: the fish gently sculls itself along with synchronous beats of the pectorals. In contrast, the pelvic fins have lost their locomotory function, becoming defensive spines. The size and arrangement of the dorsal and pelvic spines vary between species although the basic arrangement is common. Another common feature of the family is the lack of the typical teleost scales. These have been replaced by bony plates or scutes, which are greatly modified scales forming a distinct row running down each flank of the body. Lateral plates are best developed in *Spinachia*, *Gastero-steus* and *Pungitius*, vestigial in *Culaea* and absent in *Apeltes*. Inter- and intra-specific variation in the number and pattern of the lateral plates provide one of the most intriguing functional and evolutionary problems posed by the sticklebacks (Chapter 10).

Gasterosteus

Gasterosteus aculeatus (Figure 3.1). This is the best known of the stickle-back species and the one that has attracted most attention from biologists, at least partly because of the ease with which it adapts to laboratory conditions and the hardiness with which it confronts those atten-tions. Its adult size is typically between 35 and 80 mm in total length, although in a few unusual populations smaller and greater lengths occur.

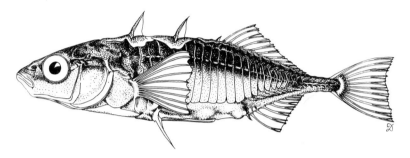

Figure 3.1: *Gasterosteus aculeatus*, a Completely Plated Morph from St. Lawrence Estuary, Canada. Total Length, 60 mm

Its common name, the threespine stickleback, is taken from the three dorsal spines that precede the dorsal fin. Each of these spines is separate from the others and the most posterior is separate from the fin, though this third spine is much shorter than the anterior two. The two pelvic spines are typically long and robust and are supported by a well developed bony pelvic skeleton, which forms a ventral shield on the belly. In some populations, however, the pelvic spines may be absent and the pelvic skeleton vestigial or absent (Chapter 10).

A feature of *G. aculeatus* is the range of variation in the develop-ment of the lateral plates (Bell, 1976; Wootton, 1976; Hagen and Moodie, 1982). Three morphological forms or morphs are recognised on the basis of the number and arrangement of the lateral plates (Figure 3.2). The completely plated morph has a continuous series of plates which run from just posterior of the head to the tail stalk, where they form a distinct caudal keel on each side of it. The partially plated morph has an anterior series of plates, then a gap before a caudal series of plates, usually with a caudal keel. Only an anterior series of plates is present in the low plated morph, the rest of the flank being naked. In some populations no plates are present. Typically, the completely

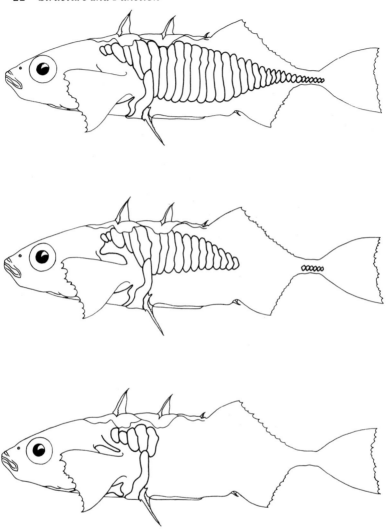

Figure 3.2: The Common Plate Morphs of *G. aculeatus*. Top, Completely Plated; Middle, Partially Plated; Bottom, Low Plated

plated morph has 30 to 35 plates in a lateral row, the low plated morph has from none to 14, and the partially plated morph has between 12 and 30. In much of the literature, the completely plated form is called *trachurus*, the low plated morph, *leiurus* and the partially plated morph, *semiarmatus*. These names do not have any formal taxonomic status but are convenient labels for the plate morphs. There is a correlation between the plate morphs and the tendency to migrate. Anadromous populations usually consist entirely or largely of the completely plated morph, although in Europe such populations often include the partially plated morph and small proportions of the low plated morph. The morph composition of freshwater resident populations is complex (Chapter 10), but populations consisting of only the low plated morph are common. The evolutionary status and significance of the three plate morphs has attracted much study (Chapter 10).

Gasterosteus wheatlandi. This geographically restricted form of *Gasterosteus* (Figure 3.3) is morphologically similar to *G. aculeatus*, but there are strong basal cusps on the pelvic spines. The number of lateral plates is variable. Fish collected from Long Island usually have plates extending to the posterior part of the body whereas fish from the Gulf of Maine tend to lack posterior plates (Perlmutter, 1963). The most conspicuous differences between the two species are in aspects of their breeding biology (Chapter 7).

Figure 3.3: *Gasterosteus wheatlandi* from St. Lawrence Estuary, Canada. Total Length, 38mm

Pungitius

Although similar in general appearance to *Gasterosteus*, *Pungitius* (Figure 3.4) is easily distinguished by the row of relatively small dorsal spines whose number ranges from seven to twelve. These spines incline alternately to one side or the other of the midline, in contrast to *Gasterosteus* in which the spines stand vertically. The body is usually relatively slim with a slender tail stalk, so that *Pungitius* looks more delicate than *Gasterosteus*. The number of lateral plates can vary from none to 34. They are usually smaller than the plates in *Gasterosteus*, but they are best developed in *P. pungitius sinensis*.

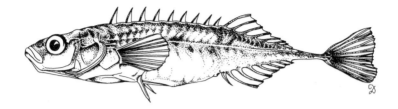

Figure 3.4: *Pungitius pungitius* from St. Lawrence Estuary, Canada. Total length, 57 mm

Culaea

In general morphology, *C. inconstans* (Figure 3.5) is similar to *Gasterosteus* and *Pungitius*, although the body is deeper than in the slender *Pungitius*. The number of dorsal spines is usually between four and six, and these are comparable in size to those of *Pungitius*. Lateral plates are present but they are so tiny that only careful study has revealed them, so on routine examination *C. inconstans* looks unplated.

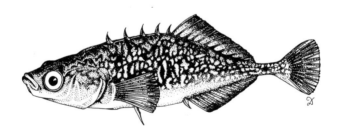

Figure 3.5: *Culaea inconstans* (Origin Not Known). Total Length, 50 mm

Apeltes

In many aspects of its breeding biology, *A. quadracus* (Figure 3.6) is an unusual stickleback (Chapter 7), but this does not extend to its general morphology. Its main features are a rather triangular body outline, an absence of any lateral plates, and between three and six relatively long dorsal spines. These dorsal spines incline to the right and left of the midline, and the last in the row is attached to the anterior of the dorsal fin.

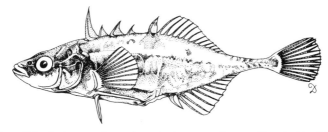

Figure 3.6: *Apeltes quadracus* from St. Lawrence Estuary, Canada. Total Length, 52mm

Spinachia

Spinachia spinachia (Figure 3.7) has a long thin body, with a long slender tail stalk. The head is also elongated, as a result of an elongation in the pre-orbital region. In the other sticklebacks, the jaw articulation lies approximately under the anterior border of the eye orbit, but in *S. spinachia* the articulation is well anterior to the eye so the elongation of the snout is not associated with an increase in the relative size of the mouth. There are about 15 spines in the dorsal row but they are small, and the pelvic spines are tiny and set well back along the body. There is a complete row of small lateral plates.

Figure 3.7: *Spinachia spinachia* from Cardigan Bay, Wales. Total Length, 95mm

The Brain and Sensory System

The brain of the stickleback reflects the relative importance of the sensory modalities in its life. Small olfactory bulbs lie on the anterior ventral surface of the telencephalon, the anterior portion of the forebrain. Most of the fibres from the olfactory bulbs project into the ventral part of the telencephalon, which consists of paired hemispheres. Behind the telencephalon, the diencephalon consists of the epithalamus, the thalamus and the ventral hypothalamus. A dorsal optic tectum forms a large part of the mid-brain, its size correlating with the importance of vision. The cerebellum, lying posterior to the optic tectum, probably controls co-ordination of movement and the maintenance of postural equilibrium. The posterior medulla oblongata merges into the spinal cord.

Behavioural and anatomical evidence suggest that vision is the dominant sensory modality. The retinal surface accounts for about 3.5 per cent of the total surface area of the body and the retina contains both rods and cones. The sensitivity of the cones to colour can be correlated with the spectral absorption of natural water. Inshore and inland waters, the characteristic habitats of sticklebacks, are green, yellow-green or orange-brown, depending on the quantities of chlorophyll and products of the decay of vegetation. In *G. aculeatus*, three types of cone are present, one sensitive to blue, one to green and one to red light; the absorption maxima for the visual pigments are 452, 529 and 604 nm, respectively. The cones are arranged in a well defined rectangular mosaic in the retina. In *Spinachia*, four types of cone are present, sensitive to blue, dark green, light green and orange (Loew and Lythgoe, 1978; Lythgoe, 1979). Behavioural studies on *G. aculeatus* have shown peaks of sensitivity to light of 510 and 594 nm in females and 502 and 594 nm in males (Cronly-Dillon and Sharma, 1968). In addition to good colour vision, sticklebacks also have good form vision (Meesters, 1940).

The epithelium of the two olfactory organs accounts for only 0.4 per cent of the surface area of the body. Although there is only a single opening to each olfactory organ, the functional morphology indicates that the flow of water over the sensory epithelium is unidirectional so that incoming and outgoing currents do not get mixed up (Theisen, 1982). Taste receptors are present in the mouth and pharynx. Olfaction seems to play little part in feeding, but olfaction and chemoreception may play a more important role in reproduction (Chapter 7).

The role, if any, of sound reception in the life of sticklebacks is not known. Sticklebacks may produce sounds with their spines (Fish, 1954), but early attempts to condition sticklebacks to sound were unsuccessful (Westerfield, 1922).

Endocrine Organs

Two types of endocrine organ can be recognised. Neuroendocrine organs represent parts of the nervous system that have become modified to secrete neurohormones; a variety of other organs form the other component of the endocrine system. At the core of the endocrine system is the pituitary organ, lying just under the brain, which acts as the primary link between the central nervous system and the endocrine system. Two components of the pituitary can be recognised. One is the neurohypophysis, which develops from an outgrowth of the floor of the brain. The function of the neurohypophysis in the stickleback is still unclear although it has been implicated in osmoregulation. A second neuroendocrine organ, the pineal complex, lies in the roof of the brain (van Veen *et al.*, 1980). This complex also contains photoreceptor cells and may be important in the control of the timing of reproduction by the photoperiod (Chapter 7). A third neuroendocrine organ, the urophysis, lies at the posterior end of the spinal cord, but again its functional significance in the stickleback is not understood.

The second part of the pituitary, the adenohypophysis, consists of several types of glandular cells whose secretions are released into the blood system and control either directly or indirectly most anabolic and catabolic processes including growth and reproduction. In the anterior part of the pituitary are the prolactin cells whose secretory product, prolactin, is important in osmoregulation for sticklebacks in fresh water (Chapter 2) and may also be important in the regulation of male parental behaviour (Chapter 7). Dorsal to the prolactin cells lie the cells that produce adrenocorticotrophic hormone (ACTH), which controls the activity of the inter-renal endocrine tissue. In the middle region of the pituitary, three types of glandular cell have been identified. One type secretes the thyroid stimulating hormone (TSH), a second secretes somatotrophic (growth) hormone (SH), and a third secretes gonadotrophic hormone (GTH), although some evidence suggests that two types of cells perhaps secreting two different GTHs may be present (Slijkhuis, 1978). The secretory activity of these cells in the pituitary is probably controlled by neurohormones produced in the brain, which either inhabit or stimulate the secretion of specific pituitary hormones. Although the general pattern of the control of the pituitary hormones in sticklebacks is similar to that of other teleosts, few of the details are known.

The major endocrine target organs for the pituitary hormones are the inter-renals, the thyroid and the gonads. Inter-renal tissue lies in the paired head kidneys and it is assumed to be homologous with the adrenal cortex tissue of terrestrial vertebrates. Thyroid tissue lies diffusely on the dorsal surface of the ventral aorta in the pharyngeal region ventral to the oesophagus. This endocrine organ has been implicated in osmoregulation and migratory behaviour in sticklebacks (Chapter 2), but probably also regulates metabolic activity. The gonads — paired testes or ovaries — lie in the abdominal cavity, dorsal to the alimentary canal. In males, the interstitial tissue is the endocrine component of the testes, producing androgens. In females, it is probable that the follicle cells which surround a maturing oocyte produce the female sex hormones, oestrogens and progesterones. Two endocrine organs which are probably concerned with the regulation of the concentrations of ions in the body fluids are the corpuscles of Stannius and the ultimobranchial organ. The former lie on the posterior dorsal surface of the kidneys, and the latter is found in the septum between the abdominal cavity and the sinus venosus just ventral to the oesophagus. Studies on the function of these organs in the stickleback are only just beginning (Chapter 2).

The endocrine component of the pancreatic tissue occurs in tissue that lies near the pyloric sphincter, which separates the stomach from the intestine. By analogy with other vertebrates, the pancreas probably regulates carbohydrate metabolism, growth and protein synthesis, although analysis of the carbohydrate content of the tissues of the stickleback suggests that carbohydrate metabolism probably forms a relatively small component of the metabolic activity. Other endocrine tissue is found in the kidneys and probably in the alimentary canal. Although the small size of the stickleback has proved an advantage in many experimental studies, it is a disadvantage in studies on endocrinology, and most of the functional significance of the endocrine organs discussed has been inferred from studies on the histological and histochemical properties of the organs.

Muscle and Movement

The trunk musculature of the sticklebacks lacks red muscle, which in teleosts is used for long periods of sustained swimming. Such red muscle operates aerobically. The trunk musculature consists entirely of white muscle which operates anaerobically (Boddeke *et al.*, 1959; Kronnie *et al.*, 1983). This white muscle is used for short rapid bursts of swimming, but an oxygen debt is built up so that the muscle fatigues rapidly in comparison with red muscle. Because they lack red muscle, sticklebacks have been classified as 'sprinters' in terms of their locomotory capabilities (Boddeke *et al.*, 1959), although this description is perhaps flattering.

Sticklebacks are not highly active swimmers. Their normal form of locomotion is a leisurely sculling with the pectoral fins — the labriform mode of locomotion (Lindsey, 1978). Associated with this mode of locomotion is a well developed pectoral skeleton. At times of excitement, when escaping from a predator, chasing a rival or approaching a female, the stickleback switches to a more typical teleost mode, carangiform locomotion. The posterior region of the body is thrown into waves which pass posteriorly to the tail. In sticklebacks, the relatively well developed post-cranial skeleton and the lateral plates tend to restrict the extent of the body that generates the locomotory wave. This carangiform swimming cannot be maintained for any extended period of time; it is simply used for short rapid bursts of swimming.

Sticklebacks spend much of their time hovering motionless in the water column. They have a well developed swim bladder which lies in the dorsal roof of the abdominal cavity. Shortly after hatching, a young stickleback fills this bladder by swimming to the surface and gulping air, but soon afterwards the connection between the bladder and the alimentary canal is lost. Thereafter, pressure changes in the gases in the bladder are controlled by the absorption and secretion of gases through its wall. The anterior portion of the bladder secretes gases into the bladder whereas the posterior portion, which is separated from the anterior portion by a diaphragm, can reabsorb them. Adjustments in the quantity of gas in the bladder allow the stickleback to maintain neutral buoyancy as the fish moves up and down in the water column (Fange, 1953).

Urinogenital System

The kidneys and gonads lie dorsally in the abdominal cavity above the alimentary canal. The paired kidneys lie in the roof of the cavity and each kidney has two distinct components, an anterior head kidney linked by a strand of tissue to the posterior or trunk portion. This is also a functional division because the head kidney contains lymphatic, haemopoietic and endocrine tissue whereas the trunk portion has, as its primary function, the production of urine (Chapter 2). This posterior portion also has an important function in reproductively active males, for the glue used in the construction of the male's nest is synthesised by the cells of the kidney tubules (Chapter 7). The gonads consist of a pair of testes or ovaries and lie just below the swim bladder. Except in young immature fish, the gonads can be distinguished macroscopically, with the ovaries soon exceeding the size of testes; the visceral peritoneum that covers the testes contains many melanophores, with their black pigment. In sexually mature females, the ovaries expand almost to fill the abdominal cavity and this causes a visible distension of the abdomen.

Digestive System

The alimentary canal forms a simple tube, with the oesophagus leading into a true stomach which is separated from the anterior intestine by a pyloric sphincter. The intestine leads to the rectum from which it is indistinguishable histologically (Hale, 1965). There are two lobes to the liver, with the larger right lobe reaching past the pyloric region of the stomach. The liver shows changes in size that are correlated both with the nutritional state of the fish and with the reproductive cycle (Chapters 6 and 7). As in other teleosts, it is probably the site at which the bulk of the material destined to form the egg yolk is synthesised (Chapter 7). Other organs associated with the alimentary canal are the pancreatic tissue and spleen. Well fed fish accumulate fat bodies around the alimentary canal, but these tend to become reduced or disappear altogether in poor feeding conditions or during winter.

Systematic Position of the Sticklebacks

Similarities in anatomy, morphology, behaviour and ecology indicate that the sticklebacks form a natural, monophyletic family, but the evolutionary relationships between the members of the family are not known. *Culaea*, *Pungitius* and *Gasterosteus* may be closely related, with *Apeltes* and *Spinachia* not closely related to that group, nor to each other. Unfortunately, the relationship of the sticklebacks to other teleosts is even more obscure. They are obviously closely related to the tube-snouts, the Aulorhynchidae, but this family contains only two species and provides no real clues to the relationships between the tube-snouts and sticklebacks and other teleosts. Conventionally, the sticklebacks are classified in the order Gasterosteiformes whose other members include fishes such as the pipefishes, seahorses and other bizarre marine forms (Wootton, 1976). In any studies of the sticklebacks it should be remembered that they represent an evolutionary sideline of the teleosts and so generalisations from sticklebacks to other teleosts should be made carefully and critically.

4 FEEDING

Introduction

Feeding behaviour determines the resources that the stickleback has available for investment in survivorship, growth and reproduction. It is the critical activity of the daily life of the fish. The details of the feeding behaviour determine both the qualitative composition of the diet and the quantity of food eaten. A feeding or foraging sequence can be broken down into a number of elements. The fish initiates a search, which eventually terminates with the discovery of a potential prey. This prey is approached and if it tries to evade capture it is pursued. If the prey does not escape, capture then ingestion follow. At each stage in this sequence after the detection of the prey, the fish may reject that prey and either turn its attention to another prey item or resume searching. The number of feeding sequences that are completed during a day will determine the quantity of food consumed, and the pattern of rejections and acceptances will determine the composition of the diet.

A detailed analysis of the feeding sequence under a range of environmental conditions (including factors such as the size distribution and density of the prey, the density of the sticklebacks, and the temperature and light regimens), which also takes account of the physiological state of the stickleback, will allow the development of a predictive model of its feeding tactics.

Detection of Prey

Sticklebacks are visual predators (Beukema, 1968; Wootton, 1976), so it is prey characteristics that provide visual cues that are important for the detection of prey. Important cues are size, movement, shape, colour contrast with the background and possibly oddity in movement, colour or shape.

The distance at which a fish can first detect a potential prey is called the reaction distance. This distance will depend on the size of the image that the prey item projects on to the fish's retina, so that larger prey can be detected at a greater distance (Eggers, 1977). In clear water, *G. aculeatus* could detect 10mm-long *Gammarus* or *Asellus* at a distance of 440mm although in turbid water the reaction distance was reduced to 260mm (Moore and Moore, 1976). A single *Tubifex* worm was detected at 150mm in every trial, and at 250-300mm in 50 per cent of the trials (Beukema, 1968). In the tests with *Tubifex*, fish with the shortest reaction distances were those that had some eye deformity such as swelling or bulging. The reaction distance to *Daphnia magna* that measured 2.8mm from the head to the base of the caudal spine was 120mm for red and 102mm for pale-yellow daphnids (Ohguchi, 1981).

Even casual observations suggest the importance of movement for the detection of prey. It is much more difficult to get a stickleback to feed on motionless food such as commercial fish food than on a bunch of wriggling *Tubifex* or enchytraeid worms. *Spinachia spinachia* preferred moving mysids to stationary ones unless the prey were moving very slowly ($0.2\,\mathrm{cms}^{-1}$). Movement of the prey increased the frequency at which the *S. spinachia* attempted or completed feeding responses (Kislalioglu and Gibson, 1976b). In an early series of experiments on the visual responses of *G. aculeatus*, Meesters (1940) found that the fish responded more to a wavy thread than to a motionless straight thread, but when the straight thread was moved, the response increased to a maximum, then declined when the thread was moved at high speeds. The speed at which the response was maximal was $3\,\mathrm{cms}^{-1}$. A similar optimum speed for the movement of prey was found for *S. spinachia* (Kislalioglu and Gibson, 1976b).

When tested with swarms of ten *Daphnia*, *G. aculeatus* directed their first attacks towards the *Daphnia* that did not match the colour of their background. The *Daphnia* were either pale-yellow or red and were tested against a background that was either pale-yellow or red (Ohguchi, 1918). This importance of the contrast between the prey and its background was demonstrated when *G. aculeatus* were allowed to hunt for single *Drosophila* larvae that were white and almost motionless. If patches of white cloth were placed on the normally dark bottom of the experimental tank, the chance that the larvae would be detected by the fish dropped sharply and the reaction distance was much reduced (Beukema, 1968). *Gasterosteus aculeatus* feeding on corixids in a pond took a greater proportion of the corixids whose colour made them conspicuous against the substrate (Popham, 1966). *Spinachia spinachia* was found to prefer dark to light mysids, although the relative importance of the stimuli presented by the mysid was such that movement was as or more important than size and both were more important than colour, and shape was least important (Kislalioglu and Gibson, 1976b).

Some evidence suggests that predators detect or select prey that differ in some respect from the majority of their kind. But in an experiment in which abnormally moving *Daphnia* were produced by amputation of parts of one or both antennae, no evidence was found to suggest that *G. aculeatus* consistently preferred the oddly moving prey (Milinski and Lowenstein, 1980). In contrast, *G. aculeatus* did turn their attention to *Daphnia* that were different in colour from other members of the swarm, even when the oddly coloured *Daphnia* were not conspicuous against the background (Ohguchi, 1978, 1981).

Foraging Behaviour in Experimental Studies on G. aculeatus

Behavioural Organisation of Foraging

If the prey is discovered on the substrate, the stickleback approaches and tilts its body, adopting a snout-down position while fixing the prey with its eyes. It seizes the prey with a rapid sideways twist and a snapping movement (Tugendhat, 1960a). High-speed photography has shown that the mouth opens with an explosive expanding movement, sucking the prey in to be grasped by the short but sharp teeth of the jaws (Figure 4.1). The snapping action is characterised by a very rapid opening of the mouth, immediately followed by a slower closure. If the prey is small, it is swallowed immediately, whereas if the prey is large it may be repeatedly grasped then spat out. This spitting action reverses the snapping because the mouth is opened relatively slowly but then closed rapidly. Since the prey are not chewed, this grasping and spitting may help to break up larger prey so that they can be easily swallowed. The speed at which the mouth opens has been found to vary with the size of the prey. Higher speeds are used when the prey is a relatively thick earthworm than when it is a slender *Tubifex*. Speeds of up to $283 \, \text{mms}^{-1}$ have been observed (Anker, 1978).

The probability that a foraging sequence ends successfully with the consumption of a prey depends on a number of factors including the hunger of the fish, the palatability and density of the prey, and the extent to which the foraging is disturbed.

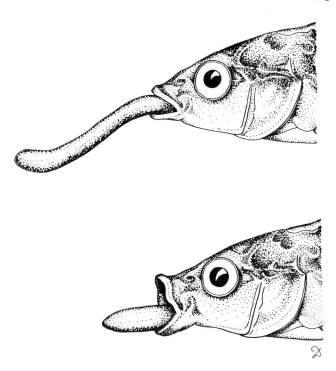

Figure 4.1: *Gasterosteus aculeatus* Feeding on a Worm. After Anker (1978)

Hunger. Typically, the hunger of sticklebacks is manipulated experimentally by depriving the fish of food for known periods of time. Two studies have analysed in detail the effect of hunger on foraging behaviour. In a series of experiments by Tugendhat (1960a, b), the sticklebacks fed on a dense population of *Tubifex* in a relatively simple feeding compartment, whereas Beukema (1968) studied sticklebacks feeding on single *Tubifex* in a relatively complex feeding compartment made up of 18 hexagonal, interconnecting cells. Despite the differences in experimental design, the basic results of these two studies were similar.

Tudendhat (1960a) found that as the period of food deprivation was increased from 1 day up to 3 days, the number of completed feeding responses during a test lasting 1 h also increased. The total number of responses that were initiated by the fish visually fixing the prey was not affected by the length of the deprivation period, but the

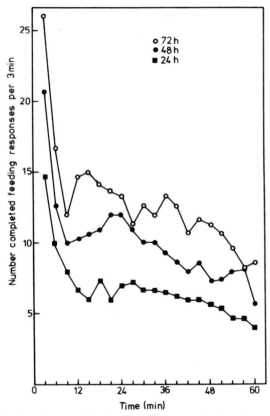

Figure 4.2: Time Course of Frequency of Completed Feeding Responses Made by *G. aculeatus* Feeding on *Tubifex* after Three Periods of Food Deprivation. After Tugendhat (1960a)

average length of time spent on a response, whether completed or just initiated, was longer the shorter the deprivation time. Although the hungrier fish completed more feeding responses, they did not spend a greater proportion of the test-period feeding. Fish spent just over 20 min making feeding responses. Within the hour-long feeding session, these effects of deprivation were reversed. As the hour progressed, the number of completed responses declined although the number of initiated responses did not. The time course of the changes in the number of completed responses during the hour was complex (Figure 4.2). There was an initial rapid decline which was fastest for the fish that had suffered the longest deprivation periods, but then there was a slight recovery in the frequency of completed responses followed by a further slow decline. Thus hunger did not influence the total time spent feeding, but increased the frequency of completed feeding responses, especially during the first part of a feeding session.

In fish that had been deprived of food for periods up to 88 h and then provided with ad lib *Tubifex* for 8 h, the total daily consumption was not related to the period of deprivation, but the proportion of the daily total that was consumed in the first hour increased with the period of deprivation (Beukema, 1968). Again, hunger affected the initial frequency of feeding rather than the final amount consumed.

When a stickleback was allowed to search for a single *Tubifex* in a complex feeding compartment, a number of quantitative components of the foraging sequence were found to increase as the period of food deprivation was increased (Beukema, 1968). Increases occurred in: the proportion of grasped prey that were eaten, the proportion of discovered prey that were grasped, the number of feeding responses directed towards inedible objects, the number of bursts of search swimming, the distance swum and the number of cells in the feeding arena visited. This positive correlation between these disparate components of foraging is good evidence that hunger can be regarded as a unitary motivational tendency which can be manipulated by food deprivation and measured by one of these covarying components. A component of foraging behaviour that was not affected by the period of deprivation was the reaction distance. As the single *Tubifex* was discovered and eaten it was replaced, so that during the course of the 40 min test period the fish became progressively satiated and the components which were positively correlated with hunger decreased in value. The total time spent handling prey was not affected by the period of deprivation because, at the shorter deprivation periods, the frequency of handling the prey decreased but the mean length of time for which each prey was handled increased.

Fish size was positively correlated with the frequency of cell visits, the proportion of prey that were grasped after discovery and the proportion of prey that were eaten after being grasped. These three components were all also dependent on the period of deprivation, which suggests that their relationship to fish size was mediated through hunger. Larger fish satiated more slowly than small fish (Beukema, 1968).

If three *Tubifex* were present rather than one, there was no effect on the number of worms eaten in the 40 min period, although there was an increase in the rate of eating in the first few minutes of the period. The main effect of the higher prey density was to reduce the proportion of prey grasped and the proportion eaten, so that each individual prey item had its risk of being eaten reduced by one-third. As the fish became more satiated, the proportion of prey grasped declined more than the proportion eaten, which suggests that the rejection of a particular prey tended to be made after it was discovered but before it was grasped (Beukema, 1968).

These results were comparable to those obtained by Tugendhat (1960a). In both series of experiments, the hunger of the experimental fish was determined by the period of food deprivation prior to the feeding session and by the number of prey consumed during the session.

Palatability. Acceptance or rejection of a prey item has a significant effect on subsequent foraging behaviour (Thomas, 1974, 1976, 1977). Sticklebacks were allowed to hunt for *Tubifex* worms in a long, linear feeding compartment. If the fish discovered a worm and ate it, then there was a reduction in the tendency of the fish to move from the area of discovery, and an immediate increase in the intensity of searching and in the frequency of snout-down approaches to the substrate. These changes in foraging behaviour led to 'area-restricted' searching. In contrast, if the *Tubifex* were rejected, there was an increased tendency for the fish to move directly away from the site of the prey and an initial decrease in the intensity of searching. This behaviour led to 'area-avoided' searching. Neither the outcome of an encounter with the prey, acceptance or rejection, nor the behaviour shown immediately after the encounter with the prey was totally dependent on the current encounter; they also depended partly on the previous feeding history of the fish. Nor could the changes in behaviour be accounted for solely in terms of an increase in the satiation of the fish during the test session. Thomas (1976) suggested that short-term positive and negative motivational after-effects were present after the acceptance or rejection of the prey, so that an acceptance positively influenced the probability that a subsequent prey would be eaten whereas a rejection negatively influenced that probability. These short-term effects of an encounter with a prey were estimated to have a time span of the order of 1-12 min. A relatively complex pattern of changes in the frequency of completed responses to prey was also seen in Tugendhat's experiment.

In Thomas's experiments, the prey was always *Tubifex*, so the cues that caused the fish to accept or reject a particular prey are not clear. Beukema (1968) analysed the effect of two prey types on the behaviour of sticklebacks foraging in his system of hexagonal cells. The new prey types used were either *Drosophila* larvae or pieces of enchytraeid worms. The densities of prey were maintained at one *Tubifex* and two of the new prey, so that two prey types were always present. The larvae were smaller than the *Tubifex* and virtually motionless. They were less acceptable to the fish than the *Tubifex*, but the rate of eating the larvae increased significantly after their introduction up to a level that was either stable or declined slowly. The presence of the *Drosophila* had no effect on the rate of feeding on *Tubifex*. In fish that found the larvae acceptable, the only component in the foraging behaviour sequence that increased consistently was the proportion of prey that was discovered. The reaction distance to the larvae did increase as the number of larvae taken increased, and there was also an increase in the number of responses to white inedible objects. Beukema suggested that the increase in the proportion of larvae discovered had two components: the fish learned fairly rapidly to respond to white objects in general and then learned, at a slower rate, to distinguish between inedible white objects and the larvae. The enchytraeid pieces were highly palatable to the fish, and nearly every piece that was grasped was eaten. As the number of enchytraeids eaten increased, the risk to the *Tubifex* declined. The increase in the number eaten was caused by an increase in the proportion that were discovered and an increase in the rate at which cells in the feeding compartment were visited. There was also an increase in the rate of response to inedible white objects, but a reduction in the rate of response to non-white inedible objects. For both types of new prey, it took many feeding experiences before the fish reached their maximum ability to discover the new prey.

In this two-prey experiment, the proportion of *Tubifex* and *Drosophila* that was eaten after being grasped declined as the fish's stomach filled, though the proportion of *Tubifex* eaten was consistently higher than the proportion of *Drosophila*. The proportion of enchytraeids eaten was high at all times. During a session, the *Drosophila* tended to be rejected earlier than the *Tubifex*, which was rejected earlier than the enchytraeid. The effect of the presence of *Drosophila* on the risk of the *Tubifex* being eaten was merely the equivalent of an increase in the density of the *Tubifex*. In contrast, the presence of the enchytraeid caused a significant reduction in the risk to the *Tubifex*. This risk fell to low levels within the first 5 min of a feeding session as the proportion of *Tubifex* discovered, grasped and eaten fell. Even in subsequent sessions in which no enchytraeids were present, the risk to the *Tubifex* did not completely recover and the fish's responsiveness to white objects remained high. The fish had learned to associate white objects with highly palatable prey (Beukema, 1968). Earlier, Meesters

(1940) had shown that sticklebacks showed a preference for snapping at drawings of solid objects after they had been eating a solid prey, whereas, after eating a thin prey, they tended to snap at drawings of lines and crosses. Thus learning the characteristics of the prey plays a significant role in the foraging behaviour of the stickleback.

The relative values of the three prey items to the sticklebacks are not known, so it is not clear whether the strong preference for enchytraieds was because they were a food of high quality.

Density of Prey. Beukema's experiments used very low densities of prey, so that at a maximum there were only three prey in an area of $187\,dm^2$. In Tugendhat's experiments the densities of *Tubifex* were much higher, but these benthic worms showed little or no evasive behaviour when approached by the stickleback. Observations on sticklebacks preying on swarms of free-swimming *Daphnia* suggest that high, local concentrations of the prey can influence the behaviour of the fish. One of the functions suggested for schooling behaviour which results in relatively high local densities of the schooling animal is that it reduces the effectiveness of a predator attacking the school and so improves the chances of each prey surviving the predatory attack (Bertram, 1978). A single stickleback attacking a swarm of *Daphnia* was observed to prey preferentially on stray *Daphnia*. This preference for strays tended to increase as the stickleback became satiated. When confronted with a dense swarm, the fish usually stopped in front of it and then often withdrew without attacking (Milinski, 1977a).

This effect of swarms of *Daphnia* on the behaviour of the stickleback was investigated by presenting the fish with different densities of *Daphnia* confined in small tubes such that the fish could attack but not ingest the prey. Immediately prior to the test, the fish had been provided with *Tubifex* ad lib for 45 min. When the fish was given the choice between pairs of tubes, one containing two *Daphnia*, the other containing up to 40 *Daphnia*, the fish showed a significant tendency to attack the two first, rather than the 40. When the choice was between two and 20, the first attacks were directed significantly more often at the 20. Other experiments showed it was the density of the swarm rather than the absolute numbers in the swarm or its volume that was the important factor. There was a critical density above which the low density (two *Daphnia*) was attacked more frequently than the higher density. As the distance between the pairs of tubes was increased, the number of first attacks directed at the two decreased although the total number of attacks on the two remained virtually constant. When the colour of the two and 40 *Daphnia* made them less conspicuous against the background, the preference for the two *Daphnia* decreased. This suggests that conspicuous stragglers are most at risk from stickleback predation. If the fish was made hungry by not feeding it for 1 day, then it preferred to attack the swarm of 40, although if the two were detected they were attacked at a high frequency. These observations were interpreted in terms of two factors. The denser swarms were more likely to be detected by the fish (the reaction distance was greater), but the denser swarm would also impose a greater confusion effect on the fish, making it more difficult for it to direct its attack at one particular prey (Milinski, 1977a).

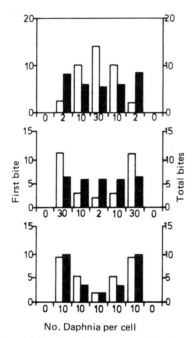

No. Daphnia per cell

Figure 4.3: Effect of Density of *Daphnia* on Number of First Responses (Open Bars) and Mean Number of Total Bites (Black Bars) by *G. aculeatus*. After Milinski (1977b)

In an extension of these experiments, the stickleback was presented with seven tubes each containing a predetermined number of *Daphnia* ranging from nil to 30 (Figure 4.3). This arrangement simulated a swarm of *Daphnia* which varied in density. The tubes were side by side, so that the fish could observe all tubes easily. The fish had not been fed for a day before the test. The first attacks were most frequently directed at the densest part of the swarm. The fish also preferred to attack the periphery of the swarm, particularly if the swarm was of equal density throughout. As the rate of attacking dropped, the fish tended to switch its attacks to the less dense regions of the swarm, although the preference for attacking the periphery was still maintained. Overall, the periphery of the swarm received more attacks than the centre, but fewer than stragglers from the swarm. Stragglers were defined as two *Daphnia* separated by one empty tube from the other tubes containing *Daphnia*. These experiments suggested that whereas a hungry fish would attack the densest part of a swarm, it would direct more and more of its attacks at stragglers from the swarm as its readiness to attack decreased (Milinski, 1977b). Experience with dense swarms did improve the performance of sticklebacks when confronted with a dense swarm (Milinski, 1979a).

The 'confusion effect' was examined with a simple but ingenious experiment in which *Daphnia* were moved over a short distance either vertically or horizontally inside a transparent tube. Speed, direction of movement and prey density were varied independently in a controlled way, so that the stimulus complexity presented to the fish could be systematically increased (Ohguchi, 1981). A stickleback tended to show a lower responsiveness to two *Daphnia* if their paths crossed at right angles than to a single *Daphnia*, although if the two *Daphnia* were not moved, the fish then showed the highest responsiveness. Attacks against the cross-moving *Daphnia* decreased as the speed of movement increased, whereas attacks against a single *Daphnia* increased with the speed of movement. The frequency of attacks was greater if the two *Daphnia* were cross-moved out of phase with each other than if the cross-movements were in phase. If the two *Daphnia* differed in colour, the frequency of attacks was higher than if they were the same colour. A fish frequently attacked a piece of carrot that was cross-moved with a red *Daphnia*, whereas if the carrot was on its own, the frequency of attacks on the carrot was low. This suggests that the fish could discriminate between the carrot and a *Daphnia* less easily when the two were cross-moving. When the experimental arrangement was altered so that the *Daphnia* could be moved vertically and parallel to each other, the frequency of attacks was higher when the two were relatively far from each other than when they were close. If the number of *Daphnia* moving vertically was increased, the frequency of attacks was decreased, although if the *Daphnia* were not moving, an increase in number increased the attack frequency. Hungry fish attacked cross-moving *Daphnia* more than satiated fish, which recalls the observation that hungry fish tended to attack the denser parts of a swarm of *Daphnia* (Milinski, 1977b).

This series of experiments suggests that a 'confusion effect' is a real phenomenon when a stickleback is attempting to attack a member of a swarm of free-swimming prey. A characteristic that makes a prey more conspicuous relative to other members of the swarm could make it more susceptible to attack, which may explain why sticklebacks tended to attack oddly coloured *Daphnia* more (page 34).

These experiments with swarms of *Daphnia* show that a hungry stickleback will initially attack the densest part of the swarm, especially if it has had prior experience of such a swarm, but turns its attention to less dense parts of the swarm and to more conspicuous members of the swarm as its readiness to attack declines. The predatory stickleback selects against a tendency for the prey to stray or to differ from other members of the swarm. Apostatic selection (see below) forms an exception.

Disturbances to Foraging. While foraging, the stickleback will be exposed to other stimuli, some of which may be unpleasant or potentially dangerous. A response by the fish to such stimuli will influence the pattern of the foraging sequences.

The effect of an electric shock on feeding behaviour was analysed by Tugendhat (1960b,c). The fish were deprived of food for 1, 2 or 3 days and exposed to a shock classified as low, medium or high when they entered the feeding compartment to start feeding on *Tubifex*, or when making a feeding response. As the intensity of shock increased, the fish spent less time feeding during the hour-long test session, but more feeding responses tended to be completed and less time was spent per feeding response. These last two are characteristic of a hungry fish. At low levels of shock, the fish completed more feeding responses than unshocked fish, but at the high levels, fewer responses were completed because of the reduction in time spent feeding. Shocked fish that returned to their home compartment during a feeding session tended to spend longer there before returning to the feeding compartment than unshocked fish.

If the fish were not shocked but prevented from feeding on the *Tubifex* by a transparent glass plate that covered the food, on subsequently being given access to the food the fish showed an increase in the time spent feeding compared with unthwarted fish. The thwarted fish tended to initiate more feeding responses by fixating a prey, but the ratio of completions was reduced. The effects of differences in the period of food deprivation prior to the test disappeared (Tugendhat, 1960c).

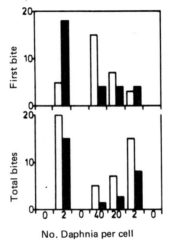

No. Daphnia per cell

Figure 4.4: Effect of Presence of Model Predator and Density of *Daphnia* on Number of First Responses and Mean Number of Attacks by *G. aculeatus*. Open Bars, Undisturbed Fish; Black Bars, Fish Disturbed by Model Kingfisher. After Milinski and Heller (1978)

The presence of a potential predator of the stickleback is likely to be a significant factor in the organisation of the stickleback's own predatory behaviour. Sticklebacks that were responding to different densities of *Daphnia* enclosed in seven adjacent cells were disturbed by being exposed to a moving model of the kingfisher, a known bird predator of the stickleback. Undisturbed fish initially attacked the highest density, but the disturbed fish preferentially attacked stragglers (Figure 4.4). Disturbed fish also attacked less frequently than the undisturbed fish (Milinski and Heller, 1978). In this situation, the disturbing stimulus produced behavioural effects similar to the effects of satiation, whereas the disturbance caused by a shock simulated the effect of hunger.

Optimal Foraging

The behavioural analysis of foraging in the stickleback leads naturally to the question of what is the functional significance of the behavioural patterns observed. The theory of optimal foraging has been developed to predict the feeding behaviour of animals, assuming that the function of foraging behaviour is to maximise the rate of food intake. It is assumed that there is a strong correlation between feeding and the Darwinian fitness of the animal, such that natural selection tends to produce optimal foragers (Krebs, 1978). Does the organisation of the foraging behaviour of the stickleback conform to the predictions of optimal foraging?

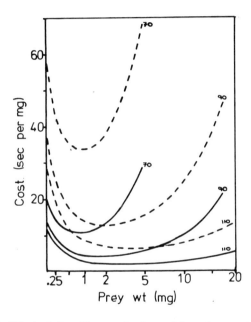

Figure 4.5: Effect of Prey Size on the Cost of Prey Capture for *S. spinachia* Feeding on Mysids. Curves Estimated for Fish 70, 90 and 110mm in Length. Continuous Line, First Prey; Broken Line, Sixth Prey. After Kislalioglu and Gibson (1976a)

In both field and laboratory observations of *S. spinachia* feeding on mysids, there was a positive correlation between prey size and fish size. Experimental studies suggested that the correlation found in the field studies arose because fish of a given size had a preferred size range of prey, and different-sized prey had different swimming speeds and escape responses. As fish became satiated, they became more size-selective (Kislalioglu and Gibson, 1975). Optimal foraging theory predicts that fish should choose the most profitable prey items, so were the *S. spinachia* selecting prey sizes that were most profitable? Cost is inversely related to profit, and the cost of taking a prey of a given size can be measured as the time spent handling the prey (h) divided by the weight of that prey (r), i.e. h/r. For *S. spinachia* feeding on mysids, the handling time was determined by the ratio of prey thickness to fish mouth size. Mouth size was, in turn, related to fish length. Handling time for a prey of a given size also increased as the fish became satiated. The cost (h/r) could be predicted for a prey of given weight for a range of fish sizes and at different levels of satiation The size of prey that minimised the cost would be the optimally sized prey for a given size of fish at a given level of satiation (Figure 4.5). When the predicted optimal prey sizes were compared with the mean prey sizes taken by fish in the field, a strong positive correlation was found (Table 4.1). The theoretical analysis suggested that larger fish would have a wider range of prey sizes that were profitable, and the field data on prey sizes did indicate that the larger fish took a wider size range of prey (Kislalioglu and Gibson, 1976a).

Table 4.1: Predicted Optimum Dimensions for Mysid Prey of *S. spinachia* Compared with Mean Prey Size in the Field (From Kislalioglu and Gibson, 1976a)

Fish length (mm)	70	80	90	100	110	120
Mouth size (mm)	2.14	2.50	2.86	3.22	3.58	3.94
Optimal prey size (mm)	7.5	9.0	10.5	12.0	13.5	15.0
Mean prey length in field (mm)	7.4	8.9	10.4	12.0	13.5	15.0

Optimal foraging theory was also used to interpret the behaviour of *G. aculeatus* feeding on *Daphnia* (Gibson, 1980). When sticklebacks were offered a choice between a pair of daphnids, one small (length 1.4 mm), one large (length 2.4 mm), the fish chose the prey that was apparently larger, irrespective of its absolute size. The smaller *Daphnia* would be less profitable so that a fish choosing on the basis of the apparent size of the prey would not necessarily be foraging optimally at all prey densities (O'Brien *et al.*, 1976). The small *Daphnia* would appear larger to the stickleback if it were closer to the fish than the large one so that the size of its image on the retina of the fish was greater. When the fish fed on mixed populations of the two size classes, they consistently over-exploited the larger, and hence more profitable, *Daphnia*. If the density of the small *Daphnia* was increased while the density of the large *Daphnia* was kept constant, the proportion of small prey taken by the fish increased. But if the densities of both large and small *Daphnia* were increased so that the relative proportion did not change, the proportion of large prey taken did not alter. At low densities of prey, the hypothesis that the fish always took the apparently larger prey predicted successfully the proportion of large and small *Daphnia* taken. At high prey densities the hypothesis predicted that a higher proportion of large *Daphnia* would be taken, but the fish did not conform to this prediction. At high *Daphnia* densities, optimal foraging theory correctly predicted the observed pattern of predation on large and small *Daphnia*. Even at the high densities, the search times for the larger prey were never short enough for the rate of food intake to be maximised by ignoring the small *Daphnia*.

Experiments with *G. aculeatus* feeding on swarms of *Daphnia* suggested that only hungry sticklebacks preferred to attack the densest portions of the swarm. As the fish became satiated they switched their attention from the denser parts of the swarm to the peripheral regions and to stragglers. Optimal foraging theory would suggest that the fish should always concentrate on the densest parts of the swarm, because the rate of consumption of prey would thus be maximised.

An important extension to the theory of optimal foraging helps to reconcile the observations and the theory (Milinski and Heller, 1978; Heller and Milinski, 1979). Their model assumes that the stickleback is confronted with two costs. The first is the cost, measured as a reduction in fitness, associated with lack of food: in motivational terms, a 'hunger' cost. The second is the cost of an increased risk of the stickleback being predated while it is concentrating on attacking prey. This cost is higher when the fish is attacking a dense swarm because the confusion effects associated with a dense swarm demand that the fish pay more attention to the task of detecting and capturing prey and so can pay less attention to the approach of a potential predator. This can be termed the 'confusion' cost. Total costs associated with foraging are the sum of the 'hunger' and 'confusion' costs. An optimally foraging fish should seek to minimise this total cost. If a fish is very hungry, it has a high 'hunger' cost which could be minimised by feeding in the densest part of the swarm. But as the hunger is diminished, the value of each additional prey is lowered because the risk of starvation is progressively lowered and so the fish can then begin to feed in less dense parts of the swarm where the 'confusion' cost is less. The model predicts that both the density of prey that the fish selects to attack and the rate of capture of prey should decline as the fish becomes satiated.

The presence of a potential predator would increase the 'confusion' cost and so should result in the threatened fish preferring to feed in less dense areas of the swarm and at a lower rate. This behaviour was indeed shown by fish exposed to a dummy kingfisher (Figure 4.4) (Milinski and Heller, 1978). The model was also relatively successful at predicting other details of the behaviour of sticklebacks attacking swarms of *Daphnia* (Heller and Milinski, 1979). The model is important because it incorporates the changes in the motivational state of the animal that are brought about by the animal's behaviour into the theory of optimal foraging.

When only one species of prey is present, optimal foraging can be used to predict, from the known profitabilities of the prey, which size classes should be taken and how the relative proportions of the size classes taken should change with alterations in prey density. An important extension to the theory is to incorporate the effects on foraging behaviour of the presence of more than one species of prey. When *G. aculeatus* were allowed to prey on *Daphnia magna* and larvae of the mayfly *Cleon dipterum*, the *Daphnia* were always preferred over the *Cleon*. The fish almost invariably caught the *Daphnia* at first attempt, but the success rate of capturing the 3-4 mm long *Cleon* was lower and decreased with an increase in the total prey density. Experiments in which the densities of *Daphnia* and *Cleon* were changed showed that the relative abundances of the two prey species were a better predictor of diet composition than the absolute density of either prey species. Although, over the densities tested, *Daphnia* was always the preferred prey, the degree of preference changed. First, the preference for *Daphnia* increased as the total prey density increased. Secondly, the preference for the relatively scarce prey increased as total prey density increased. The first result is compatible with the prediction of optimal foraging theory that as the abundance of food increases, there will be an increase in specialisation on the most profitable food items. The second result could not easily be predicted from the theory. Possibly, at high prey densities, there is a change in the profitabilities of the two prey types, perhaps as a consequence of the confusion effects created by the density of the most abundant prey. Such an effect was suggested by the reduction in the proportion of successful attacks on *Cleon* as the density of that species increased (Visser, 1982).

The detailed analysis of foraging behaviour showed that the stickleback had the behavioural repertoire to accept or reject prey in the context of prey density, type of prey and the motivational state of the fish. Further experiments have suggested that this behavioural repertoire is deployed in a manner that can be predicted in terms of optimal foraging theory, especially when this theory is extended to take account of the changing motivational state of the fish.

Apostatic Selection by *G. aculeatus*

Although the experiment with oddly coloured *Daphnia* suggested that *G. aculeatus* may select prey distinguished by some oddity from its fellows, the experiments in which sticklebacks used to feeding on *Tubifex* were then also provided with either *Drosophila* larvae or enchytraeid worms indicated the importance of learning the prey characteristics. Such learning would be easier if the new prey form was encountered relatively frequently, that is, when it was the more common prey. Such a process could lead to the more common form being taken by the predator at a greater rate than suggested by its relative representation in the total prey population, because the predator has learnt its characteristics and perhaps how to capture it. In this case the predation is directed preferentially towards the common form, not the odd form, a process called apostatic selection (Clarke, 1962).

Evidence for apostatic selection by sticklebacks came from experiments in which *G. aculeatus* were presented with populations of *Asellus* made up of a mixture of pale and dark individuals. When the population had an excess of pale over dark forms (40 pale, 10 dark), the pale forms were predated at a rate in excess of their relative abundance in the population, whereas when the dark form was the most abundant (40 dark, 10 pale), the dark form was over-predated. There was no differential predation when the two forms were present in equal numbers (Maskell *et al.*, 1977). When *G. aculeatus* were presented with varying proportions of red and pale-yellow *Daphnia*, the fish showed a preference for the red form at relative proportions that varied from 80 per cent red to 20 per cent red. When the mixture contained only 12 per cent red, the yellow form was preferred. This result was unexpected because it had been anticipated that the confusion effects would be greater at high relative densities of the red form and that the fish would switch to preying upon the uncommon yellow form. Only at low densities of *Daphnia* did the stickleback not show a preference for the red form at high proportions of the red form (Ohguchi, 1981). Because the profitability of a prey depends on its size, and on the time required for its capture and handling, the optimal foraging strategy may be either apostatic selection or selection for oddity. Selection for oddity could be optimal when capture is relatively easy but 'confusion' costs are high. If the prey is difficult to detect or capture, yet any 'confusion' cost is not high, the optimal strategy may require apostatic selection (Visser, 1981).

Foraging in Groups

Except for reproductively active males, *G. aculeatus* are frequently found in loose schools. If the fish are starved, they tend to spread apart and the school becomes even looser (Keenleyside, 1955). When one of the fish finds food, and particularly if it adopts the characteristic snout-down posture to take food off the bottom, the other fish rush towards it and start searching in the same area with several fish adopting the snout-down posture. Experiments showed that a feeding fish was a stronger stimulus to a hungry fish than either food or another fish on their own (Keenleyside, 1955). This behaviour may result in the rapid aggregation of fish in a region of high food density, with an efficient exploitation of a patchy food distribution.

The response of a group of *G. aculeatus* to a patchy distribution of food was explored by exposing six fish to two constantly renewed food patches. The food patches were formed by pipetting *Daphnia* into the two ends of the fish tank, but differed in the rate at which the *Daphnia* were added. In the first experiment, the rate of addition at one end was five times faster than at the other so the patch profitabilities were in a 5:1 ratio. The fish distributed themselves between the two patches in a ratio that closely approached a 5:1 ratio. This ratio was reached about 4 min after the start of the experiment. In a second experiment the patch profitabilities were in a ratio of 2:1 and after 9 min the profitabilities of the patches were reversed. Before the reversal, the fish again distributed themselves in a ratio that reflected the profitabilities of the patches. After the reversal, the distribution of fish approached but did not achieve the new profitability ratio. The ability of the fish to distribute themselves in ratios that reflected the density of food in the patches may represent an evolutionarily stable strategy (ESS), a foraging behaviour shown by the population which cannot be invaded by a rare mutant adopting some other feeding strategy (Milinski, 1979b, 1984).

When foraging in groups, sticklebacks may compete for the same prey, so that the profitability of a prey may partly depend on an individual stickleback's ability to capture that prey in competition with another fish. Milinski (1982) showed that, for *G. aculeatus*, successful competitors took a higher proportion of larger *Daphnia* than less successful fish. When the fish were isolated, the good competitors still preferentially attacked large *Daphnia*, but the poor competitors did not change their diet although no competing fish were present. This may result from the poor competitor learning which prey type it can successfully attack when in competition, but then retaining that preference.

The observations of the feeding of groups of sticklebacks indicate that their repertoire of foraging behaviour allows them to forage efficiently in groups as well as when they are solitary.

Foraging in Natural Environments

Laboratory experiments on foraging have used only one or two prey species and controlled prey densities, but in natural habitats the stickleback will encounter a range of potential prey species. Both the number of prey species and the overall density of prey will change seasonally and perhaps even over a day. Optimal foraging predicts that as the density of prey falls, the predator should include a wider range of prey forms in its diet. In natural habitats, a decline in prey density may also be accompanied by a reduction in the number of prey types that are available so that the stickleback is unable to increase the number of prey types that it attacks. Although the laboratory experiments have indicated that sticklebacks have the behavioural repertoire to act as optimal foragers, only the studies on *S. spinachia* have suggested that they do so in the field. There is, however, considerable information on the foraging behaviour of sticklebacks in the field (Wootton, 1976).

Habitat Differences in Diet

Although sticklebacks inhabit a wide variety of habitats, ranging from large lakes to small streams and to coastal waters, the same prey form the bulk of the diet in all these habitats. The most prominent prey types include three microcrustacean groups, the copepods, cladocerans and ostracods; the larvae and to a lesser extent the pupae of the chironomids (midges); and the aquatic nymphs of the ephemeropterans (mayflies). Of lesser importance are oligochaetes, molluscs and algae.

In brackish and coastal waters, the diet may also include larger crust-
aceans such as amphipods and isopods, and polychaete annelids (Lemmetyinen and Mankki, 1975; Manzer, 1976; Wootton, 1976; Hennig
and Zander, 1981). These larger prey are especially important for
S. spinachia (Kislalioglu and Gibson, 1975). The small size of the
sticklebacks restricts the range of potential prey types and this restric-
tion is clearly reflected in the similarity in diet over a wide range of
habitats. Where two or more species share the same habitat, the overlap
in diets can be considerable, which might lead to significant inter-
specific competition between sticklebacks for food (Chapter 8).

Seasonal Changes in Diet

As the abundance or availability of prey species changes during the
year, the diet of the stickleback changes. An example is given by
G. aculeatus living in Llyn Frongoch, a small upland reservoir in mid-
Wales. In spring and early summer, and again in late summer and
early autumn, the diet was dominated by copepods and, to a much
lesser extent, cladocerans. In mid-summer, the most important prey
were the nymphs of ephemeropterans. Chironomid larvae were taken
steadily throughout the year, whereas the pupae were taken only in
late spring and summer. Algae were eaten in autumn and winter.
Stickleback eggs were cannibalised in the late spring and summer
(Allen and Wootton, 1984). In some populations, stickleback eggs
can form a major portion of the diet in the summer (Worgan and
FitzGerald, 1981b). The seasonal changes in the diet of *G. aculeatus*
in Llyn Frongoch were similar to those of a population in a lowland
stream in north-west England (Hynes, 1950) and a small pond in
north-east England (Walkey, 1967). In the large Great Central Lake
of Vancouver Island, the diet of *G. aculeatus* was dominated by copepods and cladocerans. For two species of copepods, *Diaptomus* and
Epischura, and two species of cladocerans, *Bosmina* and *Holopedium*,
their changes in abundance in the lake were reflected in changes in
their importance to the stickleback (Manzer, 1976). This suggests
that the seasonal changes in diet are caused by changes in the density
or availability of the prey, rather than by changes in the preferences
of the sticklebacks.

In Llyn Frongoch, the stomach contents of *G. aculeatus* were
heaviest in spring and late summer and lowest in winter (Allen and
Wootton, 1984). The sticklebacks of Great Central Lake had their
highest mean weight of stomach contents in spring and early summer.
The proportion of fish with empty stomachs was lowest in spring and

early summer, highest in mid-summer and again low in the autumn (Manzer, 1976). As will be discussed in Chapter 7, spring and early summer are the breeding season, a time of high energy expenditure by the sticklebacks which must be matched by a high food intake.

Daily Changes in Feeding

As a visual predator, the foraging behaviour of the stickleback is affected by the daily cycle of light and dark. A clear example of this was shown by *G. aculeatus* in Llyn Frongoch. Samples of fish were taken at 2-hourly intervals over a 24 h period at four times in a year, once in each season. Except in May, when the night was short and moonlit, there was a decline in the weight of the stomach contents during the night, indicating that feeding had either slowed down or stopped completely (Figure 4.6)

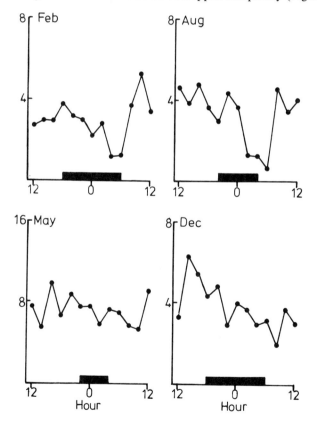

Figure 4.6: Diel Variations in Weight of Stomach Contents (mg) of *G. aculeatus* in Llyn Frongoch, mid-Wales. Black Bar Indicates Night-time. After Allen (1980)

(Allen and Wootton, 1984). In Great Central Lake *G. aculeatus* showed a similar decline during the night when sampled in July. Some of the food in the stomach from fish taken late in the night was relatively fresh, indicating that some feeding during the darkness may have occurred. There was some evidence of peaks of feeding activity in the post-dawn and pre-dusk periods (Manzer, 1976). When the stomachs of *G. aculeatus* living in a salt-marsh pool in the St. Lawrence Estuary were sampled for a 24 h period in June, the females were found to have high rates of feeding in the post-dawn period, whereas the males did not show a peak in feeding activity (Worgan and FitzGerald, 1981b). This pattern changed when weather conditions changed (G.J. FitzGerald, pers. comm.). In Llyn Frongoch sticklebacks, there was no strong evidence for a peak of feeding associated with dawn and dusk, although in February and August, the weight of stomach contents did increase rapidly after dawn. There were few changes in the overall composition of the diet in the 24 h period; few prey species were taken only in a restricted period within the 24 h. The clearest example of such a pattern was shown in May, when nymphs of neuropteran insects were taken only at dawn and at dusk (Allen and Wootton, 1984).

Experiments are needed that will define the efficiency of the foraging behaviour of sticklebacks over a range of light intensities, so that this can be related to the observed diel patterns of feeding of sticklebacks in natural populations.

Effect of Fish Size on Diet

The mouth size of the stickleback is related to its body size, so that larger fish have the potential of taking larger prey (Kislalioglu and Gibson, 1976a; Larson, 1976). The most rapid changes in body size take place in the first few weeks of the stickleback's life (Chapter 6), and this growth is associated with changes in the diet. As the recently hatched stickleback switches from endogenous feeding on its yolk to exogenous feeding on prey that it captures, the prey are usually the juvenile stages of cladocerans and copepods (Abdel'-Malek, 1968). In Great Central Lake, *G. aculeatus* less than 30 mm in length ate Rotifera, and the cladocerans *Bosmina* and *Holopedium*. Larger fish tended to feed on *Holopedium*, the copepod *Epischura*, and chironomid pupae and zooplankton eggs. Fish larvae were also taken by these larger fish (Manzer, 1976). Although these effects of size can be detected, the diet of young fish rapidly becomes similar to that of older and larger fish with only relatively minor differences.

Diet in Polymorphic Populations of Gasterosteus aculeatus

In *G. aculeatus* populations containing two or more polymorphic forms (Chapters 3 and 10), differences in diet between the various morphs may be one of the factors that allow coexistence of the morphs. In the Little Campbell River in British Columbia, the low and completely plated morphs coexist during the summer in the lower reaches of the stream. The low plated morph fed primarily off the bottom of the stream, eating the aquatic larvae of insects such as coleopterans and zygopterans and a freshwater bivalve. The completely plated morph fed on swimming or surface organisms including cladocerans and a decapod crustacean. Associated with this difference in diet was a difference between the two morphs in the mean number of gill rakers which project from the inner margins of the gill arches. The completely plated morph had more gill rakers, which were longer and finer than those of the low plated morph (Hagen, 1967). Rakers probably play a role in the feeding of fish by acting as a filtering device and are usually more numerous and finer in planktivorous fishes.

Paxton Lake on Texada Island, British Columbia, contains two morphs, the limnetics and the benthics (Larson, 1976). The benthics show a variable development of the lateral plates, pelvic girdle and pelvic spines (Bell, 1974), whereas the limnetics have between nine and fourteen lateral plates and a normally developed pelvic girdle and pelvic spines (Chapter 10). The gill rakers of the limnetic morph are more numerous than those of the benthic form, but the benthic form has a significantly wider mouth. The major food items of the benthics were the benthic amphipod, *Gammarus lacustris*, and chironomid larvae and pupae. Some benthics, mostly those less than 55 mm long, also included some zooplankton in their diet. The limnetics fed mostly on cladocerans although there was some feeding on *Gammarus*, ostracods, chironomid larvae and mayfly nymphs. Laboratory experiments indicated that the limnetics were more successful at feeding on a cladoceran, *Daphnia*, than the benthics because the former spent less time having to search for and then manipulate the cladoceran. In tests where an amphipod, *Hyalella*, was the prey, the benthics could take larger sized prey than limnetics of the same size and spent less time manipulating the amphipod (Larson, 1976).

The dietary differences between morphs noted in the Little Campbell River and Paxton Lake were associated with differences in habitat preferences, so there was a spatial as well as a dietary separation of the morphs (Chapter 10).

Rates of Food Consumption

Experimental Studies

The most important factors that influence the quantity of food consumed by a stickleback are probably its size, physiological state and the temperature. Experimental studies have sought to determine the quantitative relationships between food consumption and these factors.

Hunger. Hunger does not seem to influence the amount consumed on a daily basis. *Gasterosteus aculeatus* that had been deprived of food for periods that varied between 16 and 88 h did not differ in the weight of *Tubifex* they consumed in an 8 h period. They did differ considerably in the pattern of food intake during the 8 h period. Fish deprived for 88 h consumed 70 per cent of their total intake in the first hour, whereas fish deprived for 16 h consumed only about 35 per cent (Beukema, 1968).A similar result was obtained when sticklebacks were deprived of food for 3 days, then fed for 12 h with *Tubifex*, deprived for a further 12 h then fed for 12 h. The proportion of the daily food intake consumed in the first 2 h on the first day was significantly higher than the amount consumed in the equivalent period on the second day, although the total consumed in the two 12 h periods did not differ significantly (Cole, 1978).

Body Weight and Temperature. The relationship between the amount of food consumed and the body weight of *G. aculeatus* when *Tubifex* was supplied ad lib was described by the relationship:

$$\ln F = a + b \ln W$$

where F is weight of food, W is body weight, a and b are regression parameters and ln is logarithm to base e (Wootton *et al.*, 1980b). The value of the exponent, b, was significantly less than unity, which indicates that as the fish got heavier they consumed a relatively smaller weight of food. The same relationship between food consumption and body weight has been found in several fish species and the exponent is frequently less than unity. Ursin (1967) suggested that the maximum rate of consumption is a function of the area of the absorbing surface of the gut which is not directly proportional to body weight. The value of the parameter a in the above relationship is temperature dependent. For *G. aculeatus* feeding ad lib over a temperature range from 3 to 19°C, the relationship between food consumption, body weight and temperature was described by:

$$\ln F = a + b_1 \ln W + b_2 \ln T$$

where T is temperature and a, b_1 and b_2 are regression parameters. For example, when the food was *Tubifex*, the values of b_1 and b_2 were 0.52 and 0.66, respectively (Wootton *et al.*, 1980b). This relationship indicates a monotonic rise in the rate of food consumption over the temperature range used. In other species, as the upper lethal temperature is approached, the rate of food consumption then starts to decline (Elliott, 1981). This phenomenon probably also occurs in the stickleback. Although metabolic activities are in general temperature sensitive (Chapter 5), the causal mechanisms by which temperature influences the rate of food consumption are not clearly understood.

Physiological State. There is some unsystematic evidence that the physiological state, particularly in relation to the phase in the cycle of sexual maturation, affects the rate of food consumption. In the period between successive spawnings of the female *G. aculeatus*, a period of a few days, the rate of food consumption was initially high, but in the 24 h before spawning it declined. This decline has been used as an indication that the female is about to spawn (Wootton and Evans, 1976). At the end of the spawning season, the rate of consumption by females provided with a daily ration of food equivalent to 16 per cent of their body weight showed a decline (D.A. Fletcher and R.J. Wootton, unpublished). A study of the effect of sexual maturation on the rate of food consumption and foraging behaviour would provide a valuable insight into the interactions of the two essential systems of feeding and reproduction.

Control of Food Consumption

Assuming that sufficient food is present in the environment, the rate of food consumption will be controlled by two systems: the demands made by the rate of metabolic activity, that is, the systemic demand, and the rate at which the food can be processed by the alimentary canal. Little is known about the mechanism by which systemic demand regulates food consumption in the stickleback. That such effects exist is indicated by the effects of deprivation on the initial rates of food consumption and also by the effects of sexual maturation.

The rate at which food is processed by the alimentary canal can be measured by the rate at which the contents of the stomach are evacuated.

In common with other species of fish, the rate at which the stomach of the stickleback evacuates is proportional to the weight of the contents of the stomach. This property can be described by an exponential model of stomach evacuation:

$$dS/dt = -kS$$

where S is the weight of stomach contents and k is the rate constant (Cole, 1978). Sibly (1981) suggests that such exponential evacuation may be an optimal tactic if there is a cost associated with processing food at the maximal rate possible for the alimentary canal. The rate of evacuation of the stomach is faster at higher temperatures (Cole, 1978).

For a fish that is feeding at a rate F, the change in the weight of the stomach contents can be described by the function:

$$dS/dt = F - kS$$

This expression can be used to provide estimates for the rate of food consumption over a known time period if the change in the weight of stomach contents over that period and the value of k are known (Elliott and Persson, 1978). The pattern of food intake observed in sticklebacks that have been deprived of food for a pre-determined period and are then supplied with food ad lib can partially be explained by a model which assumes that the fish feed until the weight of stomach contents reaches a value of about 5-6 per cent of total body weight, and thereafter the rate of intake equals the rate of stomach evacuation (Beukema, 1968).

Food Consumption in Natural Populations

In laboratory experiments the rate of food consumption can be accurately measured, but the rate of food consumption in natural populations can only be estimated. Several methods have been proposed for such estimations, but all present some difficulties in application.

A widely used method assumes a balanced energy budget in which the energy of the food consumed can be equated with the energy dissipated through metabolism plus the energy stored as growth or reproductive products (Chapter 5). A simple formulation of this was provided by Winberg (1956) who proposed that food consumption could be estimated from the relationship:

$$0.8C = dB/dt + R$$

where C is food consumed, dB/dt is growth and R is total metabolism, all measured in units of energy. Winberg argued that the total metabolism, R, could be estimated as twice the standard metabolism that could be determined in laboratory studies (Chapter 5). Subsequently, Winberg's equation has been refined, for example by Majkowski and Waiwood (1981), but the original formulation is still frequently used.

Laboratory experiments yield quantitative relationships between growth and food consumption which can then be used to predict the rate of food consumption in a natural population from measurements of the latter's growth rate (Chapter 6) (Allen and Wootton, 1982a,b). Such laboratory experiments can also yield quantitative relationships between food consumption and the rate of production of faeces, which can then be used if the rate of faecal production in the natural population can be measured (Allen and Wootton, 1983).

The diel pattern of change in the weight of stomach contents can also yield estimates of the rate of food consumption, especially if the rate of stomach evacuation is known (Elliott and Persson, 1978; Eggers, 1979).

Food Consumption by Gasterosteus aculeatus in Llyn Frongoch

These methods were used to estimate the rate of food consumption of *G. aculeatus* in Llyn Frongoch (mid-Wales). This is a population of slow-growing fish which reaches a weight of about 0.5-0.6g and a length of about 35mm at the end of the first year of life.

The estimates for total food consumption for an average-sized stickleback in the period between the beginning of July and the end of the following June are shown in Table 4.2. Most of the estimates are reasonably close. The exception is the estimate based on the rate of faecal production. This discrepancy was partly the result of choosing an unrepresentative food for the experimental analysis of the relationship between faecal production and food consumption and partly caused by the difficulty of sampling faecal production in the field (Allen and Wootton, 1983).

Estimates for the changes in the rate of food consumption within the year were made for the 1977 year class based on growth rates and the energy budget. Although the estimates for the annual consumption by the methods were similar, and there were good correlations between the methods, the energy budget tended to give higher estimates during the winter months at low water temperatures, whereas

Table 4.2: Comparison of the estimates of the total annual food consumption by an average-sized stickleback in Llyn Frongoch

Method of estimation	Food consumption (mg)	Reference
Growth rate	3200	Allen and Wootten (1982b)
Energy budget		
(i) Winberg I	2300	Allen (1980)
(ii) Winberg II	4200	Allen (1980)
(iii) Majkowski and Waiwood (1981)	2700	Wootton (unpublished)
Diel samples		
(i) Elliott and Persson (1978)	3000	Allen and Wootton (1984)
(ii) Eggers (1979)	3000	Allen and Wootton (1984)
Faecal production	12800	Allen and Wootton (1983)

in late spring and summer, with higher temperatures, the growth method tended to give the higher estimates (Figure 4.7). Estimates from the diel samples using the method of Elliott and Persson (1978) showed relatively good agreement with the Winberg estimates for the same 24 h periods (Table 4.3).

Table 4.3: Estimates of rate of food consumption (mg day^{-1}) from four diel samplings and from Winberg's equations

Date	Temperature (°C)	Fish weight (mg)	Elliott and Persson	Winberg I	Winberg II
February	5.5	237	4.8	4.9	9.5
May	15.0	476	15.6	16.1	28.1
August–September	15.5	187	6.4	8.2	15.2
December	4.0	228	6.2	4.1	8.4

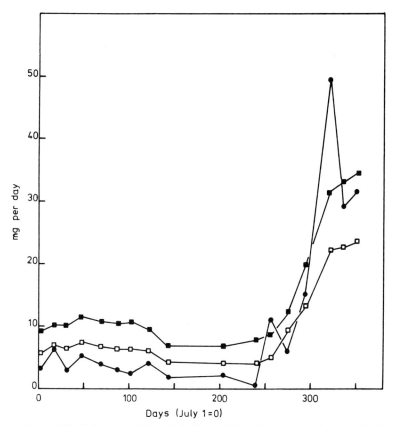

Figure 4.7: Estimates of Mean Daily Rate of Food Consumption by an Individual *G. aculeatus* in Llyn Frongoch between July and June. Closed Squares, Winberg II Estimate; Open Squares, Majkowski and Waiwood (1981) Method; Closed Circles, Allen and Wootton (1982b) Growth Equation

From diel samples Manzer (1976) estimated the daily ration of *G. aculeatus* in Great Central Lake to be 6.55 per cent of their body weight in July and 7.8 per cent in October. The daily ration of *G. aculeatus* in brackish water in the Baltic expressed in terms of the consumption by a fish weighing 1 g was estimated to be 24.2 mg at 10°C, 48.3 mg at 14°C, and 132 mg at 18°C (Rajasilta, 1980). These estimates were based on measures of the rate of digestion. On the basis of the rate of energy expenditure measured by respiration rate, the daily ration of a stickleback weighing 4.5 g was estimated to vary from 80 mg in winter to 230 mg in August (Krokhin, 1957). These estimates are relatively comparable to the estimates for equivalent conditions in Llyn Frongoch.

A major problem with these estimates of the rate of consumption in natural populations is their lack of precision. Nevertheless, they do indicate the order of magnitude of the rate at which the stickleback is exploiting its prey and the rate at which energy and nutrients are being made available for partitioning into survivorship, growth and reproduction. The pattern of this partitioning will depend on the environmental conditions that the fish is experiencing (Chapter 5).

Conclusion

Experimental analysis of the feeding behaviour of sticklebacks had led to significant advances in the understanding of foraging tactics, both in terms of the behavioural mechanisms that are involved and in providing data, which have led to important extensions to the theory of optimal foraging. The recognition that the optimal strategy will change as the motivational state of the animal changes introduces a greater realism to the theory. The study of *S. spinachia* has shown that optimal foraging theory can predict some aspects of feeding in the field. As qualitative and quantitative descriptions of the diet of natural stickleback populations improve in accuracy and precision, they will provide the empirical basis for further developments of a realistic foraging theory.

5 ENVIRONMENTAL FACTORS, METABOLISM AND ENERGETICS

Introduction

The rate at which food is consumed and the way in which energy from it is partitioned by the fish will depend on the environmental conditions. At one extreme, the abiotic environmental conditions may be so hostile that the stickleback is unable to maintain its metabolic and structural integrity and dies. At the other extreme, benign abiotic conditions may permit the fish to approach a maximisation of its lifetime production of offspring, if the biotic factors of food supply (Chapter 4), predation, parasitism and competition (Chapter 8) are favourable. The effects of abiotic environmental factors on the fish have been classified by Fry (1971) as lethal, controlling, limiting, masking or directive. Any factor may act in one or more of these ways. For the sticklebacks, the most important factors are probably temperature, salinity and oxygen.

Classification of Environmental Factors

Lethal Factors

Over the range that an abiotic environmental factor such as temperature can take, two zones can be identified. The zone of tolerance is the range over which the factor does not cause a detectable change in the survivorship of the fish. Conventionally, this is often taken to mean that the factor does not cause the death of the fish within 7 days. Outside the zone of tolerance, the metabolic processes are unable to compensate fully for the breakdown in the integrity of the fish so death follows. The zone of resistance is the range over which the factor causes the death of the fish in some measurable time period. The incipient lethal level is the value of the factor beyond which the fish can no longer live for an indefinite period. It marks the boundary between the zones of tolerance and resistance. At the outer boundary of the zone of resistance is the ultimate lethal level defining the value of the factor which causes death in a short period of time, e.g. in less than 10 min. Within the zone of resistance, an important value is the effective time, that is, the time required to bring about the lethal effect. Events within the zone of resistance are usually measured in terms of the LD_{50}, the level of the factor that causes the death of 50 per cent of the population in a defined time period such as 48 or 96 h (Fry, 1971).

The lethal level of a factor may depend on the acclimation level. This is the level of the factor the fish has experienced for a sufficiently long time to have become metabolically and physiologically adapted to that level before it encounters the potentially lethal level.

Temperature. For a few species of fish, the zones of tolerance and resistance to temperature are well defined (Elliott, 1981), but this is not yet the case for the sticklebacks. For *G. aculeatus* and *P. pungitius*, their wide geographical ranges and their extensive polymorphisms suggest that details of the zones of tolerance and resistance will vary significantly between populations. *Gasterosteus aculeatus* that were collected in brackish water on the coast of Nova Scotia had an LD_{50} of 28.8°C in tests that lasted 10 000 min after they had been acclimated at 20°C and tested in a salinity of 12 ppt. The lowest upper lethal temperature of 21.6°C was noted in fish that had been acclimated at 10°C and were tested in a salinity of 30 ppt. The upper lethal temperature was a function both of the acclimation temperature and the salinity in which the fish were tested, but not of the acclimation salinity. Upper lethal temperature was higher at the higher acclimation temperature and lower at test salinities of 0 ppt and 30 ppt than at 12 ppt at acclimation temperatures of 10°C and 20°C. The effect of test salinity was particularly marked after acclimation at 10°C. No effect was found of fish size on the upper lethal temperature (Jordan and Garside, 1972). Specimens of *G. aculeatus* collected from the Columbia River on the west coast of North America had an LD_{50} of 26°C in 8429 min after acclimation at 19°C. At 32°C, 50 per cent of the sample died within 2.3 min (Blahm and Snyder, 1975). Both these studies were carried out by transferring the fish directly from the acclimation temperature to the test temperature. The critical thermal maximum (CTM) for a plateless form of the stickleback collected in fresh water in southern California was determined by raising the water temperature at a rate of 1°C per 4 min from the acclimation temperature. The CTM was defined as the temperature at which opercular ventilatory movement stopped. For fish acclimated at 8°C, the CTM was 30.5°C, whereas after acclimation at 22.7°C, the CTM was 34.6°C. Fish size had no effect on the CTM (Feldmeth and Baskin, 1976).

No comparable data are available on the lower lethal temperatures for the sticklebacks. All the species have geographical ranges which ensure that some populations regularly experience water temperatures close to 0°C.

The rate of development of the eggs of sticklebacks is directly related to temperature (Heuts, 1956; Wootton, 1976). In *G. aculeatus*, egg survival at 25°C was poor, suggesting that this temperature is close to the lethal level for eggs. At 8°C, survival was good, but it took about 40 days for the eggs to hatch (Heuts, 1956).

Salinity. Sticklebacks are euryhaline, tolerating a wide range of salinities. Lake Techirghiol, Romania, which contains a population of partially plated *G. aculeatus*, has a salinity that reaches 80-100 ppt (Munzing, 1963).

In *G. aculeatus* the range of tolerance varies between populations and polymorphic forms. Tolerance also varies with temperature and the physiological state of the fish. Heuts (1945) used the ability of the fish to maintain a constant osmotic pressure in their body fluids to define the range of salinity tolerances for anadromous and freshwater populations of sticklebacks in Belgium, over a temperature range from 4 to 20°C. Outside the breeding season, the freshwater population had its widest range of salinity tolerance at 10°C, a range from just above full sea water (~40 ppt) to a mixture of 50 per cent distilled water and 50 per cent tap water. This range was slightly narrower at 4°C and much narrower at 20°C. (Sea water has a salinity of about 35 ppt.) The anadromous form had a tolerance range between a salinity greater than sea water (45 ppt) down to tap water at 4 and 10°C, but at 20°C the lower salinity tolerance was about one-third that of sea water and the upper tolerance just below the salinity of sea water (Figure 5.1). In the breeding season, the tolerance of lower salinities increased whereas the tolerance to higher salinities decreased in both populations.

A comparison of the salinity tolerances of the eggs of the anadromous and freshwater populations yielded a complex picture (Heuts, 1947, 1956). At 23°C, a temperature which would be unusually high for the breeding season in Belgium, the eggs of the anadromous form tolerated higher salinities than those of the freshwater form. At 19°C, both forms had a bimodal survivorship curve, but with the salinities for maximum survivorship out of phase. The freshwater form had

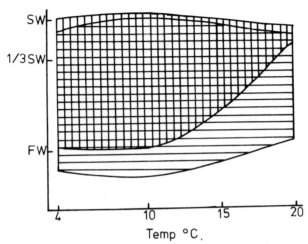

Figure 5.1: Range of Temperature and Salinity over Which *G. aculeatus* Maintains a Constant Osmotic Pressure. Salinity is on a Logarithmic Scale. Vertical Shading, Anadromous Population; Horizontal Shading, Resident Freshwater Population. After Heuts (1945)

peaks of egg survival at salinities of 0.33 ppt and 10 ppt, whereas the anadromous form had peaks of survival at 0.04 ppt and 3.3 ppt. At 10°C, the anadromous form had a single peak of survivorship at a salinity of about 3.3 ppt, and the freshwater form had a maximum survivorship at 16 ppt; the highest salinities were not tested. This series of experiments suggests, surprisingly, that the optimal salinity for egg survivorship of the freshwater form is higher than for the anadromous form over the range of temperatures likely to occur in the breeding season.

A comparative review of the salinity tolerances of other stickleback species showed that *C. inconstans* had the lowest and *S. spinachia* the highest tolerance (Nelson, 1968). *Culaea inconstans* fed actively at salinities up to 50 per cent sea water, but at 60 per cent, most feeding had stopped, and all activity stopped at 70 per cent. Death within 24 h occurred at salinities of 70 per cent and above, although the tolerance was greater at 8°C than at 16°C. Death occurred between 5 and 15 days after *C. inconstans* was placed in 40 per cent sea water. *Pungitius pungitius* from a freshwater population stopped feeding at 80 per cent sea water at 16°C and at 90 per cent at 8°C. *Apeltes quadracus*, also from a freshwater population, stopped feeding at 110 and 130 per cent sea water at 16°C and 8°C, respectively.

Even *C. inconstans* taken from an alkaline pond in North Dakota had a maximum salinity tolerance of 21 ppt (Ahokas and Duerr, 1975), which is similar to the tolerance reported by Nelson (1968) for freshwater *C. inconstans*.

Oxygen. In some circumstances, shallow bodies of fresh water may become deoxygenated. Warm, heavily vegetated waters, in which there are high rates of decomposition, and ice- and snow-covered lakes in winter can both reach low oxygen concentrations. The minimum oxygen concentration at which the stickleback can exist may be as low as 0.25-0.5 ppm, but normal active life at these levels is not possible (Jones, 1964). In tests in which 20 *G. aculeatus* were maintained in sealed 600 ml beakers, the oxygen concentration when all the sticklebacks were dead was between 3.5 and 8 per cent of air saturation. In similar tests, *P. pungitius*, which is frequently associated with denser vegetation than the threespine stickleback, had a slightly higher toleration of low oxygen concentrations (Lewis *et al.*, 1972).

Culaea inconstans can be found in lakes that become deoxygenated in winter, such as Mystery Lake, Wisconsin, where the oxygen concentration can fall to 0-1.3 mg litre^{-1}. Their ability to survive these conditions depends on their use of the microlayers of oxygen associated with gas bubbles at the interface of the ice and the water. In the absence of these bubbles, confined *C. inconstans* suffered 50 per cent mortality in 4.3 h, whereas, in the presence of bubbles, 50 per cent mortality was reached after 21.5 h. Survivorship dropped sharply as the bubbles were depleted of oxygen. As the oxygen concentration declined, *C. inconstans* had a relatively low rate of respiration and increased its ventilation rate. Its activity decreased when concentrations fell to 0.5 and 0.25 mg litre^{-1}. As the oxygen concentration of the water fell, the fish spent more time at the ice-water interface. *Culaea inconstans* was even observed to orientate to the flow of water from the operculum of a mudminnow that was pumping oxygenated water down through a crack (Klinger *et al.*, 1982).

Both *C. inconstans* and *P. pungitius* collected in southern Manitoba used the relatively well oxygenated surface film at the air-water interface when the oxygen content of the water declined at warm temperatures (Gee *et al.*, 1978).

Other Lethal Factors. Sticklebacks are sensitive to environmental pollutants such as heavy metals, industrial wastes and insecticides (Jones, 1964; Wootton, 1976), and are absent from heavily polluted waters. They will recolonise if remedial action is taken to improve water quality (Wheeler, 1979; Turnpenny and Williams, 1981). For *G. aculeatus*, water of pH less than 4.5 is frequently lethal so they are excluded from waters that become acidified through rainfall or poor land management. But, in the Outer Hebrides, populations are found in fresh water with a pH as low as 3.5. Such populations show a reduction in the development of lateral plates, the pelvic girdle and spines (Campbell, 1979; Giles, 1983a).

Controlling Factors

Controlling factors govern the metabolic rate of the fish by their influence on the state of activation of the components of the metabolic chain. A controlling factor places two bounds on the level of metabolism. It will permit a maximum rate of metabolism by its influence on the rates of chemical reactions, but it will also demand a certain minimum metabolic rate which is necessary to release the energy required for the repairs needed to keep the organism in being. The upper bound is called active metabolism and represents the maximum, sustainable rate of energy production. The lower bound is called standard metabolism and represents the minimum rate of metabolism that will maintain the steady-state integrity of the fish. It is often regarded as the metabolic rate of an unfed, resting fish. At some intermediate level is the routine rate of metabolism, which is the rate exhibited by a fish showing random, spontaneous movement. The difference between active and standard metabolism has been called the 'scope for activity' (Fry, 1971).

Temperature. This is the crucial controlling factor because of its effect on the rate of chemical reactions. Fish are ectotherms living in an environment, water, which has a high specific heat and so acts as an effective heat sink. With few exceptions, fish are isothermal with their external environment (Elliott, 1981). Sticklebacks typically live in shallow waters which can show relatively rapid changes in temperature. Although the influence of temperature on the rate of chemical reactions can be described by the Arrhenius equation such that the logarithm of the rate is proportional to the reciprocal of the absolute temperature, fish do not simply respond passively to temperature changes but are able to show considerable biochemical and physiological regulation (Hochachka and Somero, 1973).

Relatively little systematic analysis of the effect of temperature on the metabolic rate of sticklebacks has been published. When the rate respiration of *G. aculeatus* in a closed respirometer was measured at 7, 12.5 and 20°C (Cole, 1978), the relationship between the rate of oxygen consumption, body weight and temperature was described by:

$$\ln R = a + b_1 \ln W + b_2 T$$

where R is the rate of oxygen consumption, W the body weight of the fish, T the temperature, and a, b_1 and b_2 are the parameters estimated by regression. In the experiment, the values the parameters took were: $a = 2.43$, $b_1 = 0.72$, and $b_2 = 0.06$, when R was in millilitres of oxygen per hour, W in grams and T in degrees Celsius. The average routine rate of metabolism for a fish weighing 0.5 g was estimated as 0.081 ml O_2 h^{-1} at 7°C, 0.113 ml O_2 h^{-1} at 12.5°C and 0.177 ml O_2 h^{-1} at 20°C. Studies on other fish species suggest that both active and standard rates of metabolism also increase with temperature, although at high temperatures a shortage of oxygen may start to limit metabolism (Brett, 1979). The rate of food consumption also increases with a rise in temperature, at least up to some optimal temperature at which the rate is maximal (Chapter 4).

Limiting Factors

A limiting factor operates by restricting the supply of substrates or the removal of products in the metabolic chain so that the rate of metabolism is less than that permitted by the levels of the controlling factors. The crucial limiting factor for fish is most likely to be the biotic factor of food supply, but under some circumstances the supply of oxygen may be inadequate to support the rate of metabolism permitted by the environmental temperature.

Oxygen. Low oxygen concentration has been shown to restrict both activity and growth in some fish (Brett, 1979). Unplated southern Californian *G. aculeatus* showed a rapid decline in their respiration rate at 18.6°C when the oxygen concentration of the water fell below 2 ppm. Above this concentration, the rate of respiration was independent of oxygen consumption, but at concentrations below 7.4 ppm, the ventilation rate increased to maintain the respiration rate at a constant level (Feldmeth and Baskin, 1976). The energy required to maintain the high ventilation rates at the low oxygen concentrations may be diverted from growth or other activities (Feldmeth and Baskin, 1976). *Culaea inconstans* showed a decrease in locomotory activity when the oxygen concentration fell below about 1 mg litre^{-1} in water at 2.5-4.0°C (Klinger *et al.*, 1982). At 16°C, *C. inconstans* started to show a decline in the ventilation rate and in activity when the partial pressure of oxygen in the water dropped to 34.1 Torr (saturation partial pressure was approximately 150 Torr). At 18.7 Torr, 50 per cent of the fish were using the surface film. A decline in activity was shown by *P. pungitius* at 58.7 Torr, and at 49.9 Torr, 50 per cent of the fish were using the surface film (Gee *et al.*, 1978).

Masking Factors

Masking factors exert an additional metabolic demand on the fish at particular levels of the controlling and limiting factors. This metabolic demand reduces the energy that can be partitioned into other components such as growth or behavioural activity. The metabolic demand represents the energy cost the fish has to pay in order to maintain either a constant internal or external environment in the face of the masking factor (Fry, 1971).

Salinity. *Gasterosteus aculeatus* has the ability to maintain a relatively constant osmotic pressure and ionic concentration in its body fluids over a wide range of external salinities (Heuts, 1945). The rate of oxygen consumption of the completely plated and partially plated morphs was significantly higher in sea water (35 ppt salinity) than in fresh water at both 4 and 20°C, whereas the low plated form did not show a significant difference in the rate of respiration. This suggests that the osmoregulation of the completely and partially plated morphs in sea water requires the expenditure of energy whereas that of the low plated form uses a mechanism that does not require an expenditure of energy detectable by respirometry. The low plated morph osmoregulated by altering the level of free amino acids in the body fluids (Gutz, 1970). Its respiration rate showed no change over a range of salinities from 1.7 to 27.2 ppt at 12°C after acclimation for 5 weeks (S. McGibbon and R.J. Wootton, unpublished), which also suggests that salinity does not act as a masking factor for this morph. During the breeding season, the active osmoregulation of the completely and partially plated morphs became less effective and they adopted the mechanism of the low plated form. The mechanism of the low plated form was ineffective in sea water at low water temperatures (Gutz, 1970). These differences in osmoregulation are reflected in the global distributions of the three morphs (Chapter 10). Information on the effect of salinity on the growth rates of the polymorphic forms of the stickleback would be of considerable interest.

Current. If a fish wishes to maintain its position in a current, it must expend energy on mechanical work. No information is available on the effect of current as a masking factor in the stickleback, but the fish tends to choose areas of still or relatively slow-moving water and so may minimise the energy costs of maintaining spatial position.

Directive Factors

Such factors allow or require a response by the fish which is directed in relation to a gradient in the factor in space or time. These directed responses are important because they allow the fish some choice over the environment it experiences in addition to its biochemical and physiological capacities for adaptation to environmental circumstances. Sticklebacks respond to gradients in temperature and salinity.

Temperature. When *G. aculeatus* that had been collected from intertidal pools on the coast of Nova Scotia were tested in a vertical temperature gradient that ranged from 5°C at the bottom to 28°C at the top, their preferred temperature depended on the acclimation temperature and the salinity of the water (Garside *et al.*, 1977). At an acclimation temperature of 5°C, the modal preferred temperature in sea water was 13°C, but in fresh water 10°C. With acclimation at 15°C, the equivalent modes were at 17°C and 16°C, and with acclimation at 25°C, the modes were 20°C in sea water and 16°C in fresh water. The final preferendum, defined as the value at which the preferred temperature is equal to the acclimation temperature, was estimated at 16°C in fresh water and 18°C in sea water. *Gasterosteus aculeatus* collected in Oslofjorden, Norway, showed much lower temperature preferences when tested in a vertical gradient that consisted of three linked compartments which differed in temperature by 5°C. Irrespective of the acclimation temperature, which varied from 5 to 20°C, the fish preferred temperatures lower than 11°C and in most cases between 4 and 8°C (Røed, 1979). This preferred range was less than the temperatures in which the fish were naturally living for most of the year, suggesting that the result may be an experimental artefact.

Salinity. In a comprehensive series of experiments on the salinity preferences of an anadromous population of *G. aculeatus* which consisted of a mixture of the completely, partially and low plated morphs, Baggerman (1957) used a 100 cm-long trough that was divided into four compartments. Two adjacent compartments contained fresh water, the next contained water with a salinity of 7.8-8.6 ppt, and the fourth compartment had a salinity of 11.6-13.4 ppt. A thin film of fresh water connected the compartments and allowed the fish to travel between them. In this apparatus, the fish showed salinity preferences which changed during the year. In the early months of the year, there was a significant preference for fresh water, but from June onwards the preference was for salt water. These changes in preference correspond to the reproductive cycle. In late winter, the fish move from the sea to breed in fresh water, with the surviving adults and young-of-the-year moving back to the sea after the spring breeding season. The young fish showed a preference for salt water within 2 months of hatching. Experiments showed that the changes in preference were correlated with the onset of sexual maturity and the end of the breeding season, but were not dependent on the presence of developing gonads. A sharp temperature rise in November did not produce a change in the preference for salt water, whereas the same rise in the early months of the year produced a change to a preference for fresh water within a few days. Fish treated with the thyroid hormone, thyroxine, showed a change from a preference for salt water to one for fresh water within 3 to 5 days. Thiourea, an inhibitor of the thyroid, caused a preference for fresh water to change to one for salt water, again within a few days. This study indicates the close relationship between the directive response to the salinity gradient and the physiological state of the stickleback. The fish does not simply respond in a passive and non-adaptive way to the environmental conditions.

Other Factors. When presented with a choice between water at pH 6.8 and acidified water, *G. aculeatus* avoided water with a pH of less than 5.8. They also avoided water with a pH greater than about 11. Over the pH range 6.0-11.0, the fish appeared to be indifferent. When given a choice between water with a high or a low oxygen concentration, the fish crossed the boundary into the low-oxygen water without hesitation, but then became agitated and so eventually returned to the well oxygenated zone where they tended to become still. At 13°C, the escape from the poorly oxygenated zone was relatively rapid, with oxygen concentrations of less than 3 ppm. At low temperature the response was similar but took longer to develop. In this situation, the fish did not seem to recognise the presence of a gradient and only escaped from a potentially harmful situation by random movements (Jones, 1964).

The Energy Budget

Introduction

Although the pathways by which the energy intake is partitioned between outputs will depend on the environmental conditions and the physiological state of the fish, the biochemical, physiological and behavioural mechanisms that allow the stickleback to select and adapt to the environment probably maintain the fish within its zones of tolerance except in exceptional circumstances. Other causes of death, such as predation, parasitism and the costs associated with reproduction, probably far outweigh in importance the deaths caused by abiotic environmental factors.

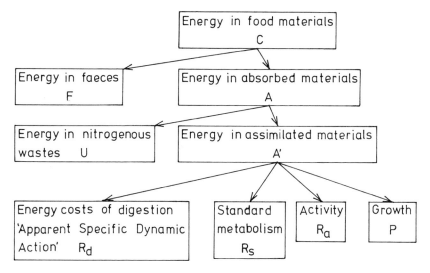

Figure 5.2: Model of Energy Partitioning by an Individual Fish. After Warren and Davis (1967)

A convenient framework for the discussion of the energy partitioning of a stickleback is the bioenergetics model of Warren and Davis (1967) shown in Figure 5.2. The energy budget for a defined time period is defined as:

$$C = F + U + P + R$$

where C is the energy content of the food consumed, F is the energy content of faeces, U is the energy content of excretory products, P

is the change in the total energy content of the body including any reproductive products released during the period, and R is the total energy of metabolism. R can be subdivided into three components, R_s, R_a and R_d, where R_s is the standard metabolism or the rate of energy expenditure of a resting unfed fish, R_a is the additional rate of energy expenditure of an active fish, and R_d is the additional rate of energy expenditure associated with the ingestion and digestion of food and is usually called the apparent specific dynamic action (SDA). A weakness of the Warren and Davis model of energy partitioning is that it does not suggest that the fish can exert control over the details of the partitioning. Models which do incorporate such control have been described by Hubbell (1971) and Calow (1973) but such models have not yet been developed for fish.

Survival, growth and reproduction do not depend only on the rate at which energy is gained and expended. They also depend on the food providing vitamins and minerals, which do not yield energy but are essential for the operation of the metabolic processes. The study of energy budgets ignores these components of the diet and so can be misleading. Unfortunately, little is known about the requirements of sticklebacks for such components (Love, 1980).

Components of the Energy Budget

Faecal Losses (F). Not all the food consumed is digested, so the fish loses some energy in the form of faeces, which will also include mucus, cells sloughed off the wall of the alimentary canal and other secretions (Elliott, 1979). Absorption efficiency is measured as the proportion of the energy in the food that is not lost as faeces. In terms of the Warren and Davis model, the efficiency is defined as: $(C - F)/C$, expressed as a percentage. Ideally, C and F must be measured in energy units, but an indication of the absorption efficiency can be obtained if they are measured in terms of dry weight. Fed to satiation twice a day with *Tubifex*, *G. aculeatus* had absorption efficiencies of between 86 and 96 per cent, where C and F were measured in energy units. The faeces had a mean energy content of 16.7 kJ g^{-1} dry weight (Cole, 1978). The absorption efficiency of fish supplied with unlimited *Tubifex* was estimated as 85 per cent and the energy content of the faeces as 18.3 kJ g^{-1} dry weight (Walkey and Meakins, 1970). In both studies, the small quantities of faeces produced by the fish meant that the energy content was measured on only a few samples. When F and C were measured in dry weights with the fish fed enchytraeid worms, the absorption efficiency was 86-96 per cent. In this experiment, absorption efficiency was higher at higher temperatures and was also higher for heavier fish, but was not related to the size of the ration consumed by the fish (Allen and Wootton, 1983). These experiments probably overestimate the absorption efficiency because any soluble components of the faeces were not measured. The natural diet of sticklebacks usually includes a high proportion of prey with chitinous exoskeletons, which are not readily digested, so the absorption efficiencies are likely to be lower. *Gasterosteus aculeatus* collected in the field produced a much higher weight of faeces in the 24 h after capture than fish produced in 24 h when fed to satiation with enchytraeid worms (Allen and Wootton, 1983).

Excretory Products (U). The stickleback also loses energy in the form of its excretory products. In fishes, ammonia is the major excretory product and each milligram of ammonia-nitrogen lost represents 24.85 J (Elliott, 1979). There has been no detailed study of the rate of production of excretory products by the stickleback, but it has been suggested that energy losses from this source can represent 3-5 per cent of the energy intake (Walkey and Meakins, 1970). Studies on the trout, *Salmo trutta*, suggest that at high temperatures and low rations such losses can reach up to 15 per cent of the intake (Elliott, 1979). Assimilation efficiency measures the proportion of the energy of the food consumed that remains after the faecal and excretory losses have been taken into account. It is defined as $[C- (F+U)]/C$. In studies on energy budgets of fish it is frequently taken to be 80 per cent, but it is probably a function of temperature, ration, composition of the food and other factors (Elliott, 1979).

The energy left after faecal and excretory losses have been taken into account can be regarded as physiologically useful energy. It is used for metabolism and stored in the form of new tissue as somatic growth and reproductive products.

Energy of Metabolism (R). The rate at which energy is being used in metabolism can be measured as the rate of heat production by the fish, or, more conveniently, by the rate of respiration. Each milligram of oxygen consumed is equivalent to an energy expenditure of about 13.56 J (Elliott and Davison, 1975). In this way the rates of oxygen consumption by the fish under various environmental conditions can be converted to rates of energy expenditure. The rate of respiration is a function of the body weight of the fish of the form:

$R = aW^b$

or in the linear form:

$\ln R = \ln a + b \ln W$

where R is the rate of respiration, W is the body weight, and a and b are parameters (cf. page 56). In fishes, the exponent, b, is typically less than unity and often approximately 0.8. For the measurement of the rate of respiration, the fish has to be enclosed in a respirometer, which typically provides only a restricted volume, so that although relatively accurate measurements can be obtained, their relevance for a free-ranging fish are more difficult to judge. The enclosed nature of the respirometer may stress the fish, so that the rates of energy expenditure are unusually high unless the fish is given time to become used to the conditions.

In a study of the effects of activity on the metabolic rate of *G. aculeatus*, single fish were enclosed in a respirometer containing 500ml of aerated water and then the respirometer was sealed and the decline in the oxygen content of the water was measured. In the first experiment, the fish were allowed to show spontaneous activity swimming at their own speed: this provided a measurement of the routine rate of metabolism. In a second experiment, the fish were made to swim against a current created by a magnetic stirrer. The speed of the current was increased until the fish could just maintain a constant position: this yielded a measurement of the maximum rate of metabolism. Eventually, the fish tired and were passively swept round by the current: this provided a measurement of the minimal rate of metabolism. The experiments were carried out at 15°C, but on fish collected in either February or August. In all conditions, the weight exponent, b, in the relationship between rate of respiration and body weight was less than unity, indicating that the heavier fish had a relatively slower rate of energy expenditure when this was expressed per gram body weight. The maximum and minimum rates were significantly higher in August than in February, but the routine rates did not differ significantly. In February, the maximum rate was 4.5 times the minimum, and in August, the ratio of maximum to minimum was 3.8 (Meakins, 1975; Meakins and Walkey, 1975). The minimum rate could be regarded as an estimate of R_s and the difference between the maximum and minimum rates as an estimate of R_a for the maximum level of activity, although the artificial environment used in the experiments makes it difficult to extrapolate these results to other situations.

A significant correlation between swimming speed and the rate of oxygen consumption was shown by *G. aculeatus* kept at 13-14°C (Lester, 1971). At a swimming speed of one fish length per second, the rate of oxygen consumption was about $1.0 \, \text{ml} \, \text{h}^{-1} \, \text{g}^{-1}$ dry weight, but this rate was doubled at twice the swimming speed.

The rate of oxygen consumption increases after the consumption of a meal up to a peak and then declines back to the pre-feeding rate, indicating an increase in the rate of energy expenditure associated with food processing. In fish this energy expenditure, the apparent specific dynamic action, typically accounts for about 9-26 per cent of the energy of the food being processed (Jobling, 1981). The causes of this increase in energy expenditure are not clearly understood. A small part represents the energy required to move the food through the alimentary canal and the processes associated with uptake of food across the wall of the gut, but the major component is probably associated with the metabolism of proteins and amino acids (Jobling, 1983b). An increase in the rate of oxygen consumption after feeding has been found in *G. aculeatus*, but the values of R_d over a range of rations and temperatures are not known (E.M. Lewis, unpublished).

Production (P). Physiologically useful energy that is not expended in metabolism is stored in the form of new tissue. Some of this new tissue increases the body mass of the fish, and some consists of reproductive products, typically eggs or sperm. In the case of the male stickleback, it also includes the glue secreted by the kidneys during nest construction. Thus, production can be divided into somatic production, P_s, and reproductive production, P_r. These two components are discussed in Chapters 6 and 7.

Energy Budgets for the Sticklebacks

It is impossible to measure simultaneously all the components of the energy budget of a fish, so the budget as a whole is not of interest. Living animals are not good subjects for a critical test of the laws of thermodynamics; rather, the laws are invoked with the assumption that the energy income will be exactly balanced by the expenditure. The important problem is to determine how the components change as environmental factors and the physiological state of the fish change. The best example of this approach is the analysis of the energy budget of the brown trout in relation to temperature, body weight and food supply (Elliott, 1979). In the stickleback, more attention has been paid to the problem of the partitioning of energy between somatic and reproductive production (Chapters 6 and 7), but some incomplete budgets have been estimated.

The effects of body weight and temperature on some components of the budget for *G. aculeatus* are shown in Table 5.1. The fish were not in reproductive condition and were fed to satiation with *Tubifex* twice a day. Regression analysis was used to obtained predictions of each of the budgetary components for selected fish weights. Several components were not estimated, including the excretory products (U), R_d, and the energy costs of activity above the routine level.

Table 5.1: Effect of Temperature and Body Weight on Some Components of the Energy Budget of *G. aculeatus* (Cole, 1978)

Temperature ($^\circ$C)	Fish weight (g)	Energy content of fish (kJ)	Food consumed (kJ day^{-1})	Food absorbed (kJ day^{-1})	Routine respiration (kJ day^{-1})	Growth (kJ day^{-1})
7	0.500	2.146	0.132	0.126	0.039	0.007
	1.000	4.494	0.212	0.200	0.053	0.020
	1.500	6.782	0.262	0.245	0.064	0.025
12.5	0.500	2.146	0.191	0.180	0.057	0.024
	1.000	4.494	0.307	0.285	0.097	0.034
	1.500	6.782	0.401	0.371	0.131	0.045
20	0.500	2.146	0.258	0.240	0.077	0.028
	1.000	4.494	0.414	0.384	0.157	0.048
	1.500	6.782	0.542	0.496	0.237	0.062

Partial energy budgets for *G. aculeatus* infected with plerocercoids of the cestode *Schistocephalus solidus* were compared with those for uninfected fish (Walkey and Meakins, 1970; Meakins, 1974a). The fish were kept at 15°C and fed an ad lib ration of *Tubifex*. Absorbed energy was estimated as 80 per cent of the energy of the consumed food, and metabolic expenditure was estimated from the rate of oxygen consumption shown by a fish swimming at the maximum rate. The average values for the components of the budget were for the unparasitised fish: assimilated energy, 0.563 kJday^{-1}; respiration, 0.242 kJday^{-1}; growth, 0.052 kJday^{-1}. In parasitised fish the average values were: assimilated energy, 0.517 kJday^{-1}; respiration, 0.293 kJday^{-1}; growth, 0.046 kJday^{-1}. For the parasitised fish, the growth component consisted of the growth of the fish and the parasites (Chapter 8).

Table 5.2: Estimates of the Components of the Energy Budget of an Average Female *G. aculeatus* from Llyn Frongoch and the River Rheidol

Population	Month	Temperature (°C)	Food consumption (kJ day^{-1})	Faecal production (kJ day^{-1})	Respiration rate (kJ day^{-1})	Growth (kJ day^{-1})
Frongoch	September	13.0	0.048	0.002	0.033	0.006
	October	9.5	0.038	0.002	0.030	0.000
	November	6.5	0.043	0.002	0.025	−0.005
	December	6.0	0.031	0.002	0.026	−0.002
	January	6.5	0.038	0.002	0.026	0.001
	February	4.0	0.045	0.002	0.024	0.024
	March	5.0	0.061	0.003	0.027	0.006
	April	7.0	0.079	0.003	0.034	0.007
Rheidol	September	12.5	0.064	0.002	0.048	0.013
	October	10.0	0.056	0.002	0.046	0.000
	November	6.0	0.044	0.002	0.037	−0.031
	December	6.5	0.063	0.002	0.036	0.017
	January	6.5	0.047	0.002	0.040	0.002
	February	5.0	0.072	0.003	0.034	0.002
	March	4.5	0.085	0.003	0.035	0.005
	April	10.0	0.147	0.004	0.054	0.020

A feature of both of these partial budgets was that a relatively high proportion of the energy income was not accounted for, indicating the problem of accurately estimating all the components of the budget. Nevertheless, the relative magnitudes of the budget components from these two independent studies are not dissimilar, which suggests that realistic estimates for at least some of these components are possible.

The estimation of energy budgets in a natural population presents problems because such estimations rely on the extrapolation of experimental studies to the natural situation. A partial budget for female *G. aculeatus* during the period from September to April was estimated for two populations in mid-Wales (Table 5.2). The rates of respiration, food consumption and faecal production were predicted from experimental studies, whereas the growth was measured by the changes in the average weight and energy content of the females during the period (Wootton *et al.*, 1978, 1980a). No estimates were available for the budget components U, R_d and R_a above routine activity, and again a significant proportion of the estimated energy income was unaccounted for, although in one case the estimated outputs did slightly exceed the estimated intake. The budget did indicate that early autumn and spring were periods of relatively high energy intake, whereas in winter the energy intake was barely sufficient for the fish to maintain the total energy content of their bodies. Until other independent estimates for the energy budgets of natural populations are produced, it is difficult to judge how realistic are those for the Frongoch and Rheidol populations.

Components of the energy budget for a population of *P. pungitius* in the arctic Lake Ikroavik, Alaska, during a 24 h period in June have also been estimated (Cameron *et al.*, 1973). Metabolic expenditure was estimated from measurements of the routine rate of respiration, which yielded the relation:

$$\ln R = -2.795 + 0.823 \ln W + 0.094T$$

where R is oxygen consumption (μl min^{-1}), W is fish weight (mg) and T is temperature ($^\circ$C). Growth was measured in the field and the energy content of the fish was determined by bomb calorimetry as 21 kJ g^{-1} dry weight. Ration was estimated from the energy balance equation:

$$0.8C = dB/dt + R$$

(see Chapter 4). The estimates are shown in Table 5.3.

Table 5.3: Estimated Energy Budget for *P. pungitius* in Lake Ikroavik for a 24 h Period in June

Mean weight (g)	Food consumption (kJday^{-1})	Rate of respiration (kJday^{-1})	Growth rate (kJday^{-1})
0.074	0.040	0.024	0.008
0.112	0.058	0.034	0.012
0.162	0.086	0.045	0.019
0.223	0.107	0.060	0.026
0.299	0.138	0.076	0.034
0.388	0.173	0.094	0.044
0.493	0.213	0.115	0.056
1.500	0.448	0.287	0.071

Conclusion

The zone of tolerance for a factor such as temperature indicates the range of that factor within which the fish can live for an extended period. The concept can be extended so that the zone of tolerance is defined for several environmental factors. An early example of this is provided by the zones of temperature and salinity within which the polymorphic forms of *G. aculeatus* can osmoregulate successfully (Heuts, 1945). This approach has a close parallel with the concept of the niche developed by ecologists, which describes the multidimensional environmental volume within which the species can live. A distinction is that the zones of tolerance describe the ranges of environmental factors within which the fish survives, whereas the niche describes the range in which the rate of reproduction is at least sufficient for the population to maintain itself. Within a zone of tolerance, there may be values of the factor which permit survival for an indefinite time but do not permit an adequate rate of food consumption for growth or for reproduction (Elliott, 1981). In the analysis of the energy budgets of the brown trout, it was found that the range of temperatures over which some growth was possible was more restricted than the zone of tolerance, and the range of temperatures over which eggs would develop was even more restricted (Elliott, 1981). Gerking (1980) has emphasised that reproduction is often the most sensitive process in relation to environmental factors, so that the environmental requirements for successful reproduction are more restricted than for survival or growth. The pattern of energy partitioning must be such that each individual fish in a population invests, on average, sufficient energy into reproductive products to at least replace itself. An average investment less than this will result in the extinction of the population in a shorter or longer time.

6 GROWTH AND PRODUCTION

Introduction

Definition of Growth

Growth is the change in size of the fish. It can be measured as growth in length, in weight or in the total energy content of the fish. Although these measures of growth are usually highly correlated, the correlations between them are not perfect. A period of linear growth need not be accompanied by a growth in weight; a change in the total energy content can take place when both weight and length are unchanged. Both weight and energy content can decline sharply, but shrinkage of the linear dimensions of the fish is slight. If growth is measured in terms of energy, then the concept of a balanced energy budget (Chapter 4) can be used to define growth as the difference between the energy assimilated and the energy expended in respiration, that is:

GROWTH = IN − OUT

(Ursin, 1979). This formulation cannot be used to define growth in weight or length, although it is easier to measure growth in these terms.

The absolute growth rate is the change in size per unit of time, so that if the unit of time is small the absolute growth rate is dW/dt, where W is the size of the fish and t is time. Absolute growth rates are often not useful for comparing growth rates under different experimental or environmental conditions because they are usually proportional to the size of the fish. A better measure is the absolute rate divided by the current size of the fish, which gives the specific growth rate (G) defined as:

$G = dW/Wdt$

The mean specific growth rate over the time interval t_1 to t_2 is given by:

$$G = \frac{\ln W_2 - \ln W_1}{t_2 - t_1}$$

where W_2 and W_1 are the sizes of the fish at times t_2 and t_1, respectively. The size of the fish at time t_2 is given by:

$$W_2 = W_1 \exp [G (t_2 - t_1)]$$

Clearly, the specific growth rate can be measured in terms of length, weight or energy, though in practice it is usually measured in terms of the length or weight. The pattern of growth shown by a fish is determined by the changes in the specific growth rate in time, so an analysis of the factors that influence the specific growth rate is essential for understanding the growth of the fish.

Relationship between Weight and Length

The relationship between these two measures of the size of a fish can be described by the function:

$$W = aL^n$$

or in the linear form by:

$$\ln W = \ln a + n \ln L$$

where W is weight, L is length, and a and n are parameters determined by regression. For a fish that grows in length and weight without changing shape or specific gravity, the exponent n takes a value of 3. A measure of the condition of the fish (K) is then given by: $K = W/L^3$. If the exponent is significantly different from 3, then a measure of the relative condition of the fish (K_n) is given by: $K_n = W/W_p$, where W_p is the weight predicted for the fish from the relationship between weight and length for the population of which the fish is a member (Le Cren, 1951).

In stickleback populations, the exponent n is typically close to 3. During a 10-year study of a population of *G. aculeatus* living in a backwater of the River Rheidol in Wales, the estimated value of the exponent for fish collected each October varied from 2.73 to 3.09. During a year, the exponent for the population of *G. aculeatus* in Llyn Frongoch varied from a high of 3.35 in July to a low of 2.87 in the following June, with the value also low during the late autumn and winter

(Allen, 1980). A population living in Priddy Pool in the Mendip Hills in south-west England that was heavily infested with *Schistocephalus solidus* had an exponent that was not significantly different from 3. In this population, the condition of the fish (K) tended to decline during autumn and winter but increased in the spring. A sharp drop in May, probably associated with spawning, was followed by a slow rise during the summer (Pennycuik, 1971a).

In *P. pungitius*, the exponent in the length-weight relationship for a population in Lake Ikroavik, Alaska, was 2.75, whereas for a population in Lake Superior, Wisconsin, it was 3. 22 (Cameron *et al*., 1973; Griswold and Smith, 1973). For a population of *P. p. sinensis* in Japan, the exponent was not significantly different from 3 (Tanaka and Hoshimo, 1979).

In some fish species, there are significant changes in the relationship between weight and length at different stages in the life cycle, so that distinct growth stanzas can be recognised. Such growth stanzas have not been reported for sticklebacks.

Relationship between Energy Content and Weight

If the energy content per unit of dry weight remains constant, then measurements of dry weight can be converted to values for total energy content directly. In practice, the energy content is often not constant. In *G. aculeatus* that were not sexually mature, the energy content of small fish was lower than for heavier fish. Fish with a mean dry weight of 0.030g had an energy content of $14.23 \mathrm{kJg^{-1}}$ dry weight, whereas the energy content of fish with a dry weight of 0.270 g was $19.87 \mathrm{kJg^{-1}}$ (Meakins, 1976). The mean energy content for fish whose wet weight ranged from 0.2 to 2.2g was $18.83 \mathrm{kJg^{-1}}$ dry weight (Cole, 1978). Sexually mature females have a higher energy content because of the enlarged ovaries, a value of $24.27 \mathrm{kJg^{-1}}$ dry weight was recorded, whereas the energy content of sexually mature males was $19.66 \mathrm{kJg^{-1}}$ (Meakins, 1976). A feature of sticklebacks is the high ash content of their bodies, a reflection of the robustness of their skeletons. In *G. aculeatus* from the River Rheidol, ash formed an average of 19.8 per cent of the dry weight (Cole, 1978), which partially accounts for the relatively low values for their energy content. Because of the variations in energy content with size, sex and physiological state, measurements of growth in terms of wet or dry weight cannot be directly converted to estimates of the growth in total energy content of the fish.

Experimental Studies on Growth

Relationship between Growth and Food Consumption

The growth rate of the fish must have some functional relationship to the rate of food intake (Parks, 1982). The problem is to determine the form of the relationship and to determine how the relationship changes with changing environmental conditions and physiological states of the fish. A number of variables that help to define the relationship are of interest (Brett, 1979). Rmax is the maximum ration that a fish can consume at a given temperature, Rmain is the ration at which the fish is just maintaining its initial size, and Ropt is the ration at which the ratio of growth to food intake is maximised (Figure 6.1).

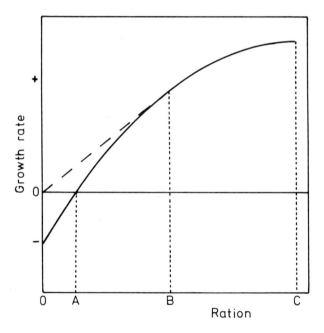

Figure 6.1: Schematic Diagram of Relationship between Growth Rate and Ration at a Given Temperature. A = Rmain; B = Ropt; C = Rmax. (Tangent from Origin to Growth Curve Which Defines Ropt Shown as Broken Line)

This ratio between growth and food intake is the growth efficiency. Three types of growth efficiency have been recognised. Gross growth efficiency (K_1) is defined as:

$$K_1 = \frac{\text{Change in size of fish in interval } t_1 \text{ to } t_2}{\text{Food consumption during interval } t_1 \text{ to } t_2}$$

A second growth efficiency is defined in terms of assimilated food, that is:

$$K_2 = \frac{\text{Change in size of fish in interval } t_1 \text{ to } t_2}{\text{Food assimilated during interval } t_1 \text{ to } t_2}$$

A third, the net growth efficiency, measures the efficiency of growth after the costs of maintaining a constant body size have been accounted for, that is:

$$K_3 = \frac{\text{Change in size of fish in interval } t_1 \text{ to } t_2}{\begin{array}{c}\text{(Food consumption during interval} - \text{Food required to}\\ \text{maintain constant body size)}\end{array}}$$

Ideally, these growth efficiencies should be calculated with both the numerator and denominator measured in energy units, but efficiencies in terms of wet weights or dry weights may be more easily obtained.

The relationship between growth and ration is typically curvilinear, with Ropt defined as the ration at which a straight line from the origin forms a tangent to the curve, and Rmain as the ration at which the curve intercepts the abscissa (Figure 6.1). The relationship between growth and ration under different environmental conditions can now be described in terms of Rmax, Ropt and Rmain and the growth efficiencies.

Temperature. The rate of food consumption and the rate of metabolism are strongly dependent on temperature in the stickleback (Chapters 4 and 5), so that temperature is a potent environmental factor affecting the relationship between growth and food consumption. In experiments with *G. aculeatus*, over the temperature range 3-19°C, Rmax for fish feeding on enchytraeid worms showed a monotonic increase (Figure 6.2), and the specific growth rate for fish eating a maximum ration also increased over the whole temperature range (Figure 6.3) (Allen and Wootton, 1982a). At 3°C, although the rate of food consumption was slow, it was still sufficient to allow a low rate of growth, and at 19°C, the highest growth rates were recorded. Studies on other species have indicated that at high temperatures the growth rate starts to decline so that the overall relationship between growth rate and

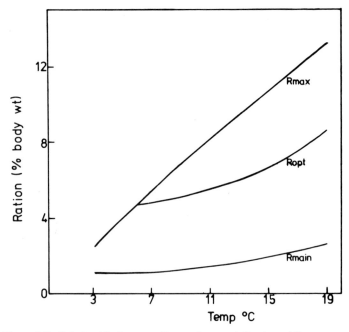

Figure 6.2: Relationship between Rmax, Ropt and Rmain and Temperature for *G. aculeatus* Feeding on Enchytraeid Worms. Simplified after Allen and Wootton (1982a)

temperature at Rmax approximately follows an inverted-U-shaped curve (Brett, 1979; Elliott, 1979). The temperature range used in the experiments with *G. aculeatus* corresponded to the range in the natural habitat, which suggests that the fish rarely experience temperatures that are so high that the growth rate is reduced.

In *G. aculeatus* the relationship between growth rate and ration was curvilinear over the temperature range 3-19°C, and at each temperature the relationship was reasonably well described by the function:

$$G = a + b \ln C$$

where G is the specific growth rate, C is the rate of food consumption expressed as a percentage of the fish's body weight and a and b are the regression parameters. G and C were measured as wet weights. This function was used to predict Ropt and Rmain. Although both Rmain and Ropt tended to increase with temperature, this increase was less rapid than the increase in Rmax (Figure 6.2). The difference

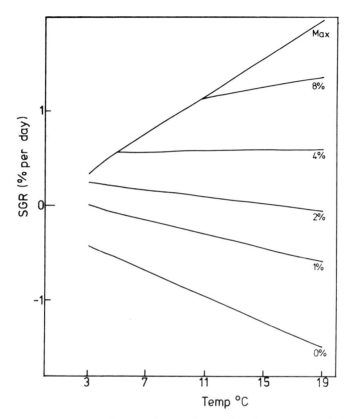

Figure 6.3: Relationship between Specific Growth Rate (Weight Per Cent), Ration as Percentage Body Wt and Temperature for *G. aculeatus* Feeding on Enchytraeid Worms. Simplified after Allen and Wootton (1982a)

between Rmax and Rmain is sometimes called the 'scope for growth', and these experiments showed that the scope for growth in the stickleback greatly increased with a rise in temperature. There was a relatively complex relationship between growth rate, ration and temperature, because at low rations the growth rate tended to decline as the temperature increased, but at high rations the growth rate increased slightly at higher temperatures (Figure 6.3). The result of this was that the highest gross growth efficiencies (K_1) were at relatively high temperatures and high rations (Figure 6.4), although the highest net growth efficiencies (K_3) were at relatively low temperatures and low rations (Allen and Wootton, 1982a). No measurements were made of the effect of ration and temperature on the energy content of the fish, so it was

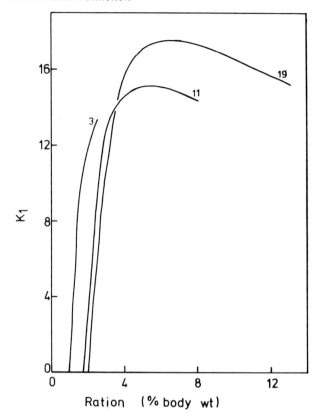

Figure 6.4: Relationship between Gross Growth Efficiency (K_1) as Percentage and Ration for *G. aculeatus* Feeding on Enchytraeid Worms at 3, 11 and 19°C. Simplified from Allen and Wootton (1982a)

not possible to estimate the growth efficiencies in terms of energy units.

Pascoe and Mattey (1977) also found a curvilinear relationship between growth rate and ration for *G. aculeatus* fed on enchytraeid worms at 15°C. They compared sticklebacks infected with *Schistocephalus solidus* with uninfected fish, but no clear differences between the two groups emerged.

Other Factors. The effects of other environmental factors on the relationship between growth and food consumption are not yet known. That such effects exist is suggested by an experiment in which *G. aculeatus* were allowed to grow under almost natural conditions until they reached a length of about 30mm and were then kept at 20°C under a photoperiod of either 16 or 8 h of light per day. Both groups received similar amounts and quality of food. The fish on the shorter photoperiod showed a more rapid growth in length than the fish on the long photoperiod. Both groups were in an early stage of gametogenesis at the start of the experiment, but in another experiment in which the fish had reached a later stage of gametogenesis, photoperiod seemed to have no effect on the growth rate (Baggerman, 1957). These results suggest that there may be complex interactions between growth, gonadal maturation and the environmental conditions, which should attract further study.

In fish, growth rate usually decreases with an increase in body size. For fish receiving ad lib rations, the relationship between specific growth rate and body weight has been described by the function:

$$\ln G = a - b \ln W$$

where G is specific growth rate, W is the body weight and a and b are the regression parameters (Jobling, 1983a). This effect has not yet been described for the stickleback, probably because most of the experiments on growth have used a restricted range of fish sizes, so that any effect of size is negligible. In a long-term study in which *G. aculeatus* were raised from hatching for periods up to 423 days, the growth rate declined as the fish grew, so that an asymptotic length was approached. This observation strongly suggests that growth rate does decline with an increase in fish size, but it is possible that the larger fish were not obtaining sufficiently high rations to support growth (van Mullem, 1967). Studies on natural populations also suggest that the growth rate of larger fish does decrease (Figure 6.6), but this effect may again be due to an inadequate food supply for the large fish rather than an intrinsic effect of body size on the growth rate.

Growth Efficiencies. By definition, the growth efficiencies at Rmain are zero, because at this ration the fish is not changing size. With rations greater than Rmain, the gross growth efficiency, K_1, of *G. aculeatus* fed on enchytraeid worms increased rapidly to a maximum and then slowly declined. The highest efficiency was reached at the highest temperature used, 19°C, and was close to 18 per cent (Figure 6.4). In this experiment both food consumption and growth were measured as wet weights (Allen and Wootton, 1982a). For *G. aculeatus* fed to satiation twice a day on *Tubifex*, the gross growth efficiency calculated in terms of energy units varied from a mean of 7.7 per cent at 7°C to 12.5 per cent at 20°C (Cole, 1978). At 15°C, *G. aculeatus* fed to satiation on *Tubifex* had a mean K_1 of 6.7 per cent if they were unparasitised and 7.5 per cent if they contained *S. solidus* (Walkey and Meakins, 1970; Meakins, 1974a). Because the growth efficiencies in the experiments with *Tubifex* as the food were calculated in terms of energy units, they are not directly comparable with those measured in the experiments in which enchytraeids were the food.

Growth of Body Organs

A problem with measuring growth in terms of the increase in the total size of the fish is that in different physiological states of the fish, the organs of the body may be showing different growth patterns. These differences in growth are obscured if only gross changes in body size are measured. In *G. aculeatus*, the effect of food on the growth of the ovaries, liver and the remaining carcase depends on the stage in the reproductive cycle. When females were kept on different rations for periods of 21 days throughout the year, the weight of the ovaries at the end of the 21 days was not usually dependent on the ration that the fish had received. The changes in the weight of the ovaries were comparable to the changes shown in the natural population and did not reflect the differences in ration. At low rations, the liver lost weight rapidly in the pre-spawning and spawning period (March to June). Even on maximum rations, the liver showed little or no growth in this period. But after spawning, the livers of fish on high rations showed high growth rates. The results suggest that, at least over a 21-day period, the growth of the ovary can be maintained even with low rates of food consumption, but at the cost of a decline in the size of the liver. The carcase may also decline in size to allow the ovaries to grow, but for most of the year the size of the ovaries relative to that of the carcase is small so the effect on the carcase is small (Allen and Wootton, 1982c). The causal mechanisms by which the relative growth rates of the body organs are regulated in relation to the rate of food consumption and other environmental factors are not understood, but such an understanding is essential for a satisfactory theory of the growth of fishes.

Models of Growth

The experimental analysis of growth and food consumption yields empirical relationships which are useful in the prediction of growth or food consumption in natural populations. However, they are unsatisfactory because without knowing the causal mechanisms that produce the observed results, the empirical relationships can only be used confidently to make predictions for the restricted range of environmental factors used in the experiments, though in natural populations a much wider range of factors will be impinging on the fish. This problem would be overcome if an adequate general model of fish growth could be developed, and a variety of models have been suggested.

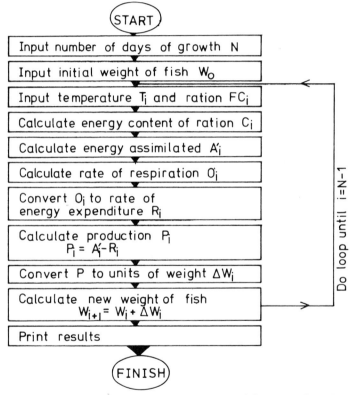

Figure 6.5: Flow Diagram for Computer Simulation of Growth in *G. aculeatus* Using a Simple Bioenergetics Model

One family of models assumes that the specific growth rate, G, depends at any instant of time on the size of the fish at that time, that is:

$G = f(W)$

Various specific functions have been suggested for $f(W)$, such as the logistic, the Gompertz and the Richard's function (Ricker, 1979). These can be useful for providing an empirical description of the growth pattern shown by a fish, but they are not acceptable as general models of growth because they do not indicate that growth rate must be a function of the rate of food consumption (Parks, 1982). The same objection can be raised against the much-used Bertalanffy model of fish growth (Pitcher and Hart, 1982), which, although deduced using physiological assumptions, does not directly relate growth to food consumption. A general model must be based on the truism that in energy terms:

GROWTH = INCOME − METABOLIC EXPENDITURE

so both income and expenditure must be modelled in relation to the size the fish has already attained and the environmental conditions, especially temperature and food supply (Ursin, 1979). A crude model of this form for the stickleback is shown in Figure 6.5, but the satisfactory development of such models requires much more information on the causal mechanisms of growth and its regulation.

Growth in Natural Populations of *Gasterosteus aculeatus*

A similar general pattern of growth was shown in all the populations that have been studied. After hatching, there was an initial rapid increase in length, but the growth rate slowed so that an asymptotic length was approached. This general pattern of growth was modelled by the function:

$L = 41.7t^{0.288}$

where L is length in millimetres, and t is age of fish in years, for a population living in Priddy Pool in the Mendip Hills (Pennycuik, 1971a). The growth in weight followed a similar pattern, but the approach to asymptotic weight is slower. This pattern was modelled for the Priddy Pool population by the function:

$W = 0.514t^{0.79}$

where *W* is weight in grams, and *t* age in years.

The details of this general pattern vary considerably between populations, which is to be expected given the wide geographical range and type of habitat for the species.

Growth in the First Year

Three studies have described the growth in the first year of life of *G. aculeatus* in detail. In each study, samples were taken from a population at regular intervals throughout a year. It was assumed that the population was sampled at random with respect to size and that the mortality of fish of a given age was not dependent on their size, so changes in the mean size of fish in the successive samples indicated the growth pattern in the population. The populations came from three distinct habitats, but they shared a common characteristic because in all three populations few fish survived for more than 1 year. One population lived on the coast of The Netherlands. It was anadromous, moving from its overwintering habitat in the sea into the brackish water of the ditches of the island of Tholen to breed in the spring, with the young fish and any surviving adults migrating back to the sea in the autumn. This population was polymorphic, containing all three plate morphs (van Mullem and van der Vlugt, 1964). Two resident populations of the low plated morph living in hard-water streams, Devil's Brook and Bere Stream, in southern England (Mann, 1971) can be compared with a population of the low plated morph living in a small, soft-water, upland lake, Llyn Frongoch, in mid-Wales (Allen, 1980; Allen and Wootton, 1982b).

The fish in the anadromous population grew rapidly in length from the time of hatching in the ditches in April and May until they migrated through the drainage system to the sea in autumn. Growth during the winter was slow, but by the following April a mean length of about 60 mm was achieved. In contrast, the mean lengths in April for fish in Devil's Brook and Bere Stream were 44.0 mm and 43.0 mm, respectively, and in Llyn Frongoch only about 30 mm (Figure 6.6). The growth in length in the first summer and autumn of life was relatively slow in the Frongoch population, and virtually ceased between October and March.

Growth in weight was not described for the anadromous population. In the hard-water streams, there was a tendency for the autumn or winter to be a time of relatively slow growth in weight but some growth did occur, whereas in Llyn Frongoch, growth ceased during autumn and winter and over some periods a loss in weight occurred

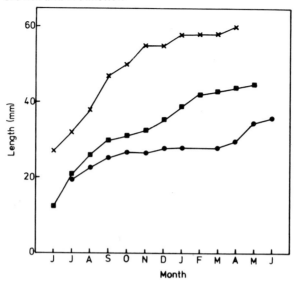

Figure 6.6: Growth in Mean Length in First Year of Life for Three Populations of *G. aculeatus*. Crosses, Anadromous Population from Tholen Island, The Netherlands; Squares, Resident Population in Devil's Brook, southern England; Circles, Resident Population in Llyn Frongoch, Wales

(Figure 6.7). There was a rapid increase in weight in the spring. Estimates of maximum rates of growth that could be achieved by the Frongoch fish were obtained from laboratory experiments (Allen and Wootton, 1982a). These estimates indicated that in late summer, autumn and winter, the fish were growing at a much slower rate than their potential, and even in spring and early summer, faster growth rates were possible (Allen and Wootton, 1982b). This suggests that the slow growth both in length and weight of the sticklebacks in Llyn Frongoch is a reflection of a poor food supply. The winter and early spring water temperatures were also lower than those of the southern hard-water streams and this would be another factor tending to reduce the growth rate in Llyn Frongoch.

Growth in Longer-lived Populations

The pattern of growth in populations of *G. aculeatus* in which the maximum life span is more than a year has not been studied in such detail, but the picture is of growth towards, though not usually reaching, an asymptotic size. Typically, growth is rapid in the first summer of life but then slows up in successive years. In two Alaskan lakes, Bare and Karluk, most of the increase in size took place in the summer

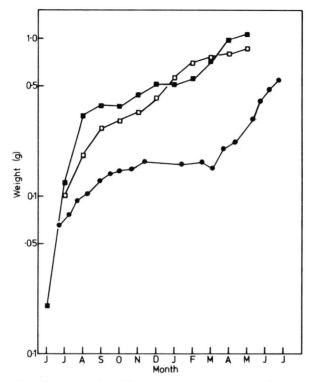

Figure 6.7: Growth in Mean Weight (on a Logarithmic Scale) in First Year of Life in Three Resident Freshwater Populations of *G. aculeatus*. Open Squares, Devil's Brook, southern England; Closed Squares, Bere Stream, southern England; Circles, Llyn Frongoch, Wales

between June and September. In Bare Lake, by the end of the first summer, the fish reached a typical length range of 38 to 42 mm; by the end of the second summer, the range was 51 to 53 mm; and the few fish that survived to the end of the third summer had a range of 55 to 73 mm. The average lengths in Lake Karluk were 39, 61, and 73-79 mm (Greenbank and Nelson, 1959). In Priddy Pool (south-west England) the oldest fish were 4 years old and their mean length was about 60 mm and mean weight only about 1.5 g (Pennycuik, 1971a). Even slower growth was shown in a population in the River Birket in north-west England, where at 1 year old the fish had an average length of 42.7 mm, at 2 years 45.6 mm, and at 3 years 51.4 mm (Jones and Hynes, 1950). Unusually high growth rates seem to be shown by sticklebacks in Lake Washington near Seattle for they reach a mean weight of about

1.8g by their first December (Eggers *et al.*, 1978). The causal factors for these differences in growth patterns are not known; there may be genetic differences between populations and major differences in the quality and quantity of food available.

Growth in First Month

Growth of *G. aculeatus* in the first month or so after hatching is difficult to measure in natural populations because the small size of the young fish makes them difficult to sample. In Llyn Frongoch, young fish from a single nest had reached a mean length of 14.8 mm and a mean weight of 0.031 g within 32 days of first being observed (Allen and Wootton, 1982b). In laboratory cultures, fish reached a mean length of 17.3 mm 29 days after hatching at 20°C (van Mullem, 1967). The specific growth rates were particularly high in this first month of life, both in the natural population and in the laboratory cultures. For the Frongoch fish, the average specific growth rate over the 32 days was 0.030 mm day^{-1} and 0.095 g day^{-1}, whereas for the laboratory population the mean specific growth rate in terms of length over the 29 days was 0.037 mm day^{-1}.

Growth in Other Species

Only the growth patterns of *P. pungitius* have been described in any detail. In the River Birket, *P. pungitius* coexisted with *G. aculeatus* and in the first few months of life had a higher growth rate than the latter species. But after the first year growth almost ceased when a mean length of about 40 mm was reached (Jones and Hynes, 1950). In Lake Ikroavik, Alaska, the population of *P. pungitius* reached a mean length of 22 mm by the end of the first August after birth, and by the end of the second and third August, the mean lengths were 42 and 68 mm, respectively (Cameron *et al.*, 1973). Females in a population in Lake Superior, Wisconsin, reached a mean length of 80 mm at 5 years old; the oldest males had an average length of about 68 mm at 3 years (Griswold and Smith, 1973). When fish from this population were reared in the laboratory at 12°C, they had a mean length of 5.65 mm at hatching and reached a mean length of 12.91 mm after 22 days. This was a mean specific growth rate of 0.038 mm day^{-1} (Griswold and Smith, 1972).

Growth of Body Organs

The overall changes in the total weight or energy content of the fish reflect changes in the weight and energy content of the body organs.

The ovaries of a female stickleback during sexual maturation in the spring show the most dramatic changes, but throughout the year the relative growth rates of different organs will vary. The growth in dry weight of the ovaries, liver and carcase of an average female *G. aculeatus* from the Llyn Frongoch population and from the River Rheidol demonstrated this clearly. All three components showed a rapid growth in late summer, reflecting that of the fish in the first few months of life. In autumn and winter, the carcase and liver stopped growing but the ovaries maintained a slow positive growth. In spring, growth of all three components again increased, the upsurge in growth of the carcase tending to precede that of the liver and ovaries (Wootton *et al.*, 1978). The observation that during the winter the ovaries maintained some growth is consistent with the result from the experimental studies, which showed that the growth of ovaries was relatively insulated from differences in food supply, whereas the growth of the carcase and especially the liver was sensitive to it (Allen and Wootton, 1982c). Well fed sticklebacks tend to accumulate fat stores around their alimentary canal. The females of Frongoch and the Rheidol showed a decrease in the size of this store during the late autumn and winter, with some recovery in spring, just prior to the breeding season. In the Rheidol population, the fat stores disappeared during the breeding season (G.W. Evans and R.J. Wootton, unpublished).

Energetics and Efficiency of Growth

The energy content of the carcases of female *G. aculeatus* from Frongoch and the Rheidol fluctuated during the year. The highest values, about $21\,kJg^{-1}$ dry weight, were reached in early autumn and again in the spring just before the breeding season. They declined to a minimum in midwinter and again during the breeding season. Both populations showed a similar pattern, although the recovery from the winter minimum was earlier in the Rheidol population (Wootton *et al.*, 1978).

Sufficient information on the growth and food consumption of the 1977 year class of sticklebacks in Llyn Frongoch was available for calculation of the gross growth efficiency in terms of wet weights. Over the whole year from July to July, it was estimated that on average each fish consumed 3.15g of food while growing from a weight of 0.066 to 0.552g, with an average gross efficiency of 15.4 per cent. The highest efficiencies were achieved in late summer and early autumn and again in the spring. In late autumn and winter the efficiencies were usually low or could not be calculated because the fish lost weight (Allen and Wootton, 1982b). The gross growth efficiencies for the population of *P. pungitius* in Lake Ikroavik were calculated for a single June day using the balanced energy equation (Chapter 5). The estimated mean efficiency was 24.7 per cent (Cameron *et al.*, 1973). Unfortunately, the two sets of figures are not directly comparable because one set was estimated in units of weight and the other in energy units.

Production in Natural Populations

The new flesh that is synthesised by the growing fish in a population forms the production of that population. This production is potentially available to predators of the population, but if not exploited by the predators, then it becomes available to the decomposers as the fish die. Total production in a population is defined as:

$$P = A - R$$

where P is production, A is the total energy assimilated by the population over a given time period, and R is the total energy respired by the population in that time period. Production is measured over a defined time period, typically 1 year. Total production has two components, production in the form of somatic growth, and production in the form of reproductive products. The somatic production by a population can be measured as:

$$P_s = G\bar{B}$$

where P_s is somatic production over the time interval dt, G is the mean specific growth rate over that time period, and $\bar{B}$ is the mean biomass present during dt. P_s is measured in terms of wet or dry weights or energy units per unit time.

Two methods are available for estimating P_s in a natural population over 1 year. The first is Allen's graphical method, in which for each month (or shorter or longer time intervals), the number of survivors of each age class in the population is plotted against the mean weight of fish in that age class. The production by the age class during the year is given by the area under the curve (Mann, 1971). A second method assumes that the mortality in the population can be described by an exponential model with a constant and known instantaneous mortality coefficient, Z, so that:

$$N_t = N_0 e^{-Zt}$$

where N_t and N_0 are numbers in the population at times t and 0. With this assumption P_s is defined as (Pitcher and Hart, 1982):

$$P = \frac{GB (e^{G-Z} - 1)}{G - Z}$$

Both these methods have been applied to populations of *G. aculeatus*. Mann (1971) used Allen's method to estimate the production in Bere Stream as 19 kg ha^{-1} year^{-1} and in Devil's Brook as 6 kg ha^{-1} year^{-1}. Because of difficulties in accurately estimating the abundance of the sticklebacks in weedy sections of the streams, these are probably underestimates of stickleback production. In both streams, stickleback production represented only a small proportion of total fish production. The second method was used to estimate production by *G. aculeatus* in several small streams in Poland (Penczak, 1981; Penczak *et al.*, 1982). Production ranged from 35.1 kg ha^{-1} year^{-1} to 0.8 kg ha^{-1} year^{-1}. In most of the streams, production by *G. aculeatus* accounted for less than 10 per cent of the total fish production, though they tended to be the most abundant fish. Three of the Polish streams also contained *P. pungitius*, and in two of these streams the production of the latter species was much higher. It reached 84.7 kg ha^{-1} year^{-1} in a stream in which production by *G. aculeatus* was only 5.0 kg ha^{-1} year^{-1}. In Lake Washington, *G. aculeatus* production was estimated at 0.39 kg ha^{-1} year^{-1} and represented only 0.4 per cent of total fish production (Eggers *et al.*, 1978). Production by *G. aculeatus* in Lake Dalnee, USSR, was estimated at 2.2 kcal m^{-2} year^{-1}, which is approximately 0.2 kg ha^{-1} year^{-1} (Krogius *et al.*, 1972). In these two lakes, *G. aculeatus* forms part of the fish community feeding on zooplankton. All these estimates suffer from the difficulty that the number of recruits to the first age class, that is, the number of fry that hatched, was not known. These preliminary estimates do suggest that, in a variety of habitats, sticklebacks account for only a small proportion of the total fish production.

Conclusion

The growth of the somatic component of the fish builds the framework that supports the process of reproduction (Miller, 1979a). In sticklebacks, the growth of this framework is relatively rapid in the first few months of life. But in the populations so far studied, with the onset of low temperatures in autumn and winter, the reproductive system (at least in the females) has a priority over the somatic component. Spring is characterised by a surge in growth of both the somatic and reproductive components, with the former preceding the latter in time. The experimental studies suggested that at low temperatures there is only small scope for growth, but at higher temperatures typical of late spring and summer the scope for growth is great. Over most of the normal temperature range for the stickleback, a higher rate of food consumption leads to higher growth. Even at high rates of food consumption, the gross growth efficiencies are relatively high compared with the maximum efficiency that can be achieved at the optimal ration. Fish which, through their foraging behaviour, can achieve high rates of food consumption will grow more rapidly than less successful foragers. This growth is partly represented by an increase in the dimensions of the body [which may be an advantage in the presence of predators (Chapter 8)], partly by an increase in the energy reserves stored in the body, and partly by an increase in the size of the gonads, especially in the females. The relationship between growth and food consumption provides a system by which natural selection can improve the foraging strategy of the fish. Food supply also directly influences reproduction and this is considered in Chapter 7.

7 REPRODUCTION

Introduction

The biological success of an individual depends finally on that individual's effectiveness at 'mapping' the energy and materials of its food into progeny (Figure 1.1). Reproduction is at the core of the adaptation of the individual to its environment. In gonochoristic organisms such as the sticklebacks, successful reproduction requires the co-operation of another animal so that the haploid gametes can be fertilised. Other individuals may also be seeking the co-operation of that sexual partner, so if partners are a relatively scarce resource, there may be competition for them. Much of the reproductive behaviour of sticklebacks can be interpreted in the context of competition for and the choice of sexual partners.

Sticklebacks provide valuable material for the study of the causation and adaptive significance of patterns of reproductive biology because they provide comparative studies at several levels. The first level of comparison is that between the sexes. In sticklebacks the reproductive investment of the female is primarily cytoplasmic in the form of eggs. During a breeding season, the female can produce, at intervals of a few days, a weight of eggs that is a significant portion of her total body weight. Relatively little time is spent in courtship behaviour so the bulk of her time is available for foraging for food. In contrast, the reproductive investment of the male is primarily behavioural. The male becomes territorial during the breeding season, defending an area against conspecifics and potential predators of the eggs and fry. He builds and cares for the nest and assumes the parental duties. The testes remain small, even during the breeding season, and the reproductive biology of the males makes it unlikely that copious amounts of sperm are produced. A second level of comparison is between populations of the same species. Recent research has shown that there can be considerable inter-population variation in the details of the reproductive biology within a species. A third level of comparison is between the stickleback species. Although the general pattern of

103

reproduction is similar in all of them, the differences that do exist provide a series of intriguing puzzles.

In their reproductive biology, the sticklebacks must solve a series of problems, the most important of which are: when to reproduce, where to reproduce, which sexual partners to choose, how much time and energy to devote to reproduction, how many propagules to produce, and what size of propagule.

Timing of Reproduction

Breeding Season

In northern fresh- and brackish-water fish communities, there is a relatively narrow 'window' during the year within which it is adaptive to produce free-living young (Wootton, 1984). This is probably determined by the seasonal availability of food in adequate quantities and of a suitable size spectrum for the recently hatched young. Most species ensure that their young are produced within this 'window' by timing their reproduction to coincide with it. The breeding season of the stickleback fits this pattern. Typically, breeding takes place at some time in the period between late March and early August, although in a few populations there may be some breeding through to the end of August (van Mullem, 1967). Where several species are sympatric, there is usually considerable overlap in their breeding seasons. In the St. Lawrence Estuary, *G. aculeatus*, *G. wheatlandi* and *P. pungitius* were breeding in the same salt-marsh pools between May and the end of June, whereas *A. quadracus* was breeding in tidal creeks close by up to the end of July (Worgan and FitzGerald, 1981a). Further south, on Long Island, *G. aculeatus* and *G. wheatlandi* bred between March and May. The breeding season of *G. wheatlandi* tended to be later but the overlap between the species was considerable (Rowland, 1983a). In contrast, in the Little Campbell River in British Columbia, the low plated resident freshwater population of *G. aculeatus* bred from March through to June, whereas the anadromous completely plated population in the same river bred between May and September, so the overlap in breeding seasons was relatively small (Hagen, 1967).

In *G. aculeatus*, the breeding season tends to be later in the more northerly populations. In Italy and Spain, breeding may be in March and April, whereas in the Baltic region, Scandinavia and Greenland, breeding is in May to late July (Bertin, 1925; Borg, 1982a).

Control of Timing

If sticklebacks are to avoid breeding at an unsuitable time of the year, they must have some mechanism that accurately brings them into reproductive condition at the appropriate season. Sticklebacks live in environments in which the photoperiod and temperature show strong seasonal patterns and so are suitable to be used as cues for the control of the timing of reproduction. The effect of temperature and photoperiod has been studied most intensively in *G. aculeatus* (Baggerman, 1980; Borg, 1982a).

Normally, sticklebacks become sexually mature during spring at a time of increasing day-lengths and water temperatures, although day-lengths and temperatures in spring tend to be similar to those in autumn, an inappropriate time for breeding. Collected in autumn and kept on a long 16L-8D photoperiod at 20°C, *G. aculeatus* became sexually mature within a 60-day period, so they do have the potential of reaching sexual maturity in autumn. When the autumn fish were exposed to a short 8L-16D regimen, few if any reached sexual maturity; thus the short day-lengths and low temperatures of winter inhibit the onset of maturity. The response to photoperiod seems to be all or nothing: a fish either becomes mature within about 45 days or does not become mature at all. During the course of winter and spring, the length of the light phase that is required for maturation steadily decreases, so that by spring a photoperiod of 8L-16D at 20°C is sufficient. At these short photoperiods, the length of the breeding season is curtailed, drastically for the females. The ability of fish to respond to a long photoperiod is enhanced if they have previously been exposed to a period of short day-lengths and low temperatures, such as the fish would experience naturally in winter. Maturation at long photoperiods is accelerated at high temperatures (20°C), but does occur at temperatures down to about 10°C (Baggerman, 1957, 1980).

It is not the absolute length of the period of light that is crucial in allowing maturation, but the distribution of light in the 24 h period. A high proportion of fish that were collected in autumn became sexually mature when exposed to a photoperiod of 6L-8D-2L-10D at 20°C. During late winter and early spring, experimental photoperiods such as 6L-4D-2L-12D or 6L-2D-2L-14D were sufficient to allow sexual maturation. An interpretation of these results is that sticklebacks show a daily rhythm of photosensitivity, and light will only stimulate sexual maturation if it coincides with a period of photosensitivity. In autumn and early winter, the period of photosensitivity does not start until about 14 h after daybreak, whereas in spring, it starts at 8 to 10 h after daybreak. With this mechanism, the fish are not stimulated to start maturation by the relatively long day-lengths during autumn, but the process of maturation can begin early in the year when photoperiods are still relatively short. The decline in the length of time after daybreak before the start of the photosensitive period seems to be accelerated by low temperatures (Baggerman, 1980). There are two exceptions to the general rule that sticklebacks can be brought to sexual maturity by exposure to long photoperiods and high temperatures. In the first month or so of life, fish do not respond to long photoperiods, and after a breeding season there is a refractory period in which the adult fish are insensitive to long photoperiods (Baggerman, 1957).

Although sticklebacks have good vision, the eyes may not be essential for the photoperiodic control of reproduction. Males that had been collected in November, then blinded in December, were exposed to photoperiods of either 16L-8D or 8L-16D. After a month, the blinded fish on the long photoperiod showed a significantly advanced development of secondary sexual characteristics compared with the fish on the short photoperiod (Borg, 1982b). The pineal gland contains photoreceptors (Chapter 3) and this may be the site of light reception for the photoperiodic response (van Veen *et al.*, 1980). It contains the precursors of melatonin, an indolamine, which in other vertebrates inhibits gonadal maturation. High doses of melatonin had an inhibitory effect on the maturation of the gonads of both male and female sticklebacks held at long photoperiods in November and January (Borg and Ekstrom, 1981). Melatonin may be synthesised by the pineal during the dark and may form part of an inhibitory pathway acting on the hormonal system that controls sexual maturation.

In common with other vertebrates, the gonads of sticklebacks are controlled by gonadotrophic hormone(s) secreted by the pituitary, which in turn is controlled by the hypothalamic region of the brain. Release of gonadotrophin is stimulated by a gonadotrophin-releasing-factor (GtRF), which has been localised in several areas of the brain of sexually mature sticklebacks (Borg *et al.*, 1982). The result of the interaction of the seasonally varying abiotic factors of light and temperature with the neuroendocrine system of the stickleback is a precise timing of the breeding season at an ecologically appropriate time.

Gametogenesis

A major aspect of the timing of reproduction is the control of the process of gametogenesis. At the right time of the year, the females must have ripe eggs that can be spawned and the males must have viable sperm, and be competent to fertilise the eggs. The difference in the patterns of investment of male and female sticklebacks is reflected in their patterns of gametogenesis, which have been described most fully in *G. aculeatus* (Baggerman, 1980; Borg, 1982a; Borg and van Veen, 1982) and in *C. inconstans* (Braekevelt and McMillan, 1967; Ruby and McMillan, 1970).

Oogenesis and the Growth of the Ovaries

The pattern of maturation of the ova of sticklebacks follows the general teleostean pattern (Scott, 1979). Germ cells, the oogonia, initially divide by normal mitotic division to give the oocytes. These then enter a meiotic division, which when completed will have reduced the chromosome number in the ova to the haploid condition. The oocytes enlarge and their nuclei contain numerous nucleoli. Each oocyte becomes surrounded by an envelope of cells consisting of an outer thecal layer and an inner granulosa layer. An oocyte, together with its cellular envelope, forms a follicle. Between the oocyte and the granulosa layer, a vitelline membrane develops and this eventually forms the outer shell of the spawned egg. The nucleoli disperse and the next event is the formation of vacuoles in the cytoplasm of the oocyte, initially in the periphery but eventually filling the cytoplasm. These vacuoles are sometimes known as the primary yolk, but are believed to eventually form the cortical alveoli which play a part in the events that follow fertilisation of an ovum (Wallace and Selman, 1979). The next phase is vitellogenesis, the deposition of the true yolk. Initially this is deposited in granules in the periphery of the cell, but the granules fuse together until yolk almost completely fills the ovum, with the cytoplasm forming a thin film on the periphery. Once vitellogenesis is completed, the maturational phase begins. The oocyte hydrates and the nucleus, now called the germinal vesicle, breaks down. The follicle ruptures and the ovum is released into the fluid-filled ovarian cavity ready to be spawned, leaving behind the post-ovulatory corpora atretica. Under some circumstances, the normal maturation of oocytes during the vitellogenic phase is arrested and the oocytes are resorbed, forming, during the process of resorption, pre-ovulatory corpora atretica. The development of all the oocytes is not synchronous so various stages will be present together in an ovary at a given time. However, a batch of oocytes does develop in synchrony to form a clutch of eggs. Ovaries can be classified in terms of the latest stage of oocyte maturation represented in the ovary.

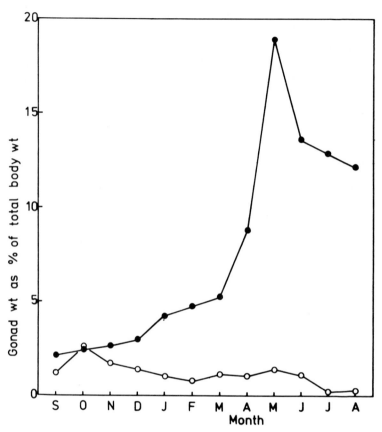

Figure 7.1: Changes in Mean Relative Weight (Gonadosomatic Index) of the Ovaries (Closed Circles) and Testes (Open Circles) of *G. aculeatus* from River Rheidol, Wales, during Annual Reproductive Cycle

Oogenesis starts in late summer and by winter some oocytes contain vesicles. Development up to this stage takes place even if the fish are kept on a short photoperiod, but the second phase including vitello-genesis is stimulated by long photoperiods. In autumn, the ovaries of *G. aculeatus* form only 2-3 per cent of the total body weight; they grow relatively slowly until the following spring by which time they represent about 5 per cent of body weight (Figure 7.1). The ovaries then grow rapidly so that just before the first spawning they form 20-30 per cent of the body weight, and the abdomen of the female bulges. Immediately after spawning they have collapsed back to 6-8 per cent but within a few days have increased back to their maximum relative size (Figure 7.2) (Wootton, 1974a; Wootton *et al.*, 1978; Borg and van Veen, 1982). The slow increase in size of the ovaries over the winter seems to be maintained even when other components of the body such as the carcase and liver are losing weight (Wootton *et al.*, 1978). Feeding experiments suggested that the ovaries were relatively insensitive to the rate of food consumption (Chapter 6) (Allen and Wootton, 1982c).

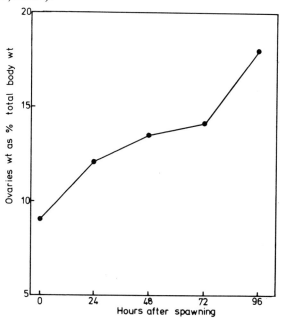

Figure 7.2: Changes in Mean Relative Weight of Ovaries in 96 h Following Spawning, Showing Rapid Increase in Size of Ovaries

Inter-spawning Interval. During the breeding season a well fed female may produce 15 to 20 clutches, a total weight and volume of eggs that far exceeds the storage capacity of the abdominal cavity, so that some phases of oogenesis must continue throughout the breeding period. Each spawning can be seen as culmination of a cycle that began with the preceding spawning, with the period between the successive spawnings defined as the inter-spawning interval (Wootton, 1974a; Wallace and Selman, 1979). During an inter-spawning interval, a population of follicles with diameters greater than 1.1 mm completes the final stage of maturation including germinal vesicle breakdown and hydration, and a second population of follicles becomes vitellogenic. A third group consists of small primordial follicles with a range of diameters up to about 0.5-0.6 mm, the largest of which are in the earliest phase of vitellogenesis. The final maturation and the recruitment of a population of vitellogenic follicles may be linked, but up to ovulation the final maturation is more rapid than the phase of vitellogenesis so the ovulated eggs are spawned while the population of follicles destined to form the next clutch is still undergoing yolk accumulation. There is a considerable increase in the weight of the ovaries, part of which is caused by an increase in the water content, but a significant component is other materials, primarily yolk. It is probable that the precursor of the yolk, vitellogenin, is synthesised in the liver and circulates in the bloodstream from which it is sequestered by the vitellogenic follicles. Thus most of the yolk has an exogenous origin. The liver can be stimulated to synthesise vitellogenin by oestrogens which are synthesised in the ovaries under the influence of gonadotrophin (Ollevier and Covens, 1982). Gonadotrophin and steroids produced by the ovaries probably also control the final maturational stages that lead up to ovulation into the ovarian cavity and spawning. A feature of the inter-spawning interval is the rate at which the ovaries grow, indicating very high rates of synthesis of the yolk precursors in this period (Wootton, 1974a).

Ovarian Regression. Even under constant environmental conditions of long photoperiods and warm temperatures, a female eventually stops spawning. Any remaining vitellogenic ova may be resorbed and the ovaries revert to the state of early oogenesis, with only oogonia and the early stages of the oocytes present. This ovarian regression occurs more rapidly at high temperatures, which in some populations may induce such regression. The final stages of follicle maturation are temperature sensitive: they are inhibited at temperatures much above 21°C and extremely slow at temperatures below 11°C (Wallace and Selman, 1979; Borg and van Veen, 1982). It is not known what causes the regression if temperatures are not too high. It is possible that the supply of oocytes that can be recruited to vitellogenesis becomes exhausted, or the neuroendocrine system controlling reproduction becomes exhausted at this stage.

Most females will eventually spawn a clutch of ovulated eggs even if a male is not present, but occasionally the eggs are not spawned but are retained in the ovarian cavity, a condition called 'over-ripeness'. An over-ripe female has a swollen abdomen, but this is relatively hard to the touch and the outlines of the eggs can be seen. There is little fluid in the ovarian cavity in contrast to the normal condition for ovulated eggs. The follicular cells, which remain after the ovulation of the eggs to form a post-ovulatory corpora atretica, may be responsible for secreting a steroid which stimulates the ovarian epithelium to produce the ovarian fluid required to maintain the ovulated eggs in good condition (Lam *et al.*, 1978, 1979). Over-ripeness seems to be a pathological condition rather than part of the normal process of oogenesis and ovulation.

Spermatogenesis and Secondary Sex Characteristics of the Male

The sexual maturation in the male falls into two clear phases. The first is the process of spermatogenesis by which mature spermatozoa capable of fertilising eggs are produced. The second is the maturation of cells in the testes which can synthesise steroid sex hormones, particularly androgens such as testosterone. Most of the secondary sexual characteristics of the male are androgen dependent.

Spermatogenesis begins in late summer and continues through the autumn so that by winter the testes contain viable, functional sperm. The germ cells, or spermatogonia, which lie in the walls of the tubules of the testes, divide mitotically to give the primary spermatocytes. These form clusters of cells contained in sacs in the tubules. Meiotic division of the spermatocytes gives the secondary spermatocytes, which by the second meiotic division yield the haploid spermatids. The walls of the sac break down, releasing the spermatids into the lumen of the tubules where they mature into the functional spermatozoa (Scott, 1979; Borg, 1982a). This phase is usually reached by winter and there is little further change until the second phase starts in late winter. During the second phase, the interstitial or Leydig cells increase in activity, secreting androgens which stimulate a number of the secondary sexual characteristics of the male stickleback. It is this second phase which is stimulated by long photoperiods. Active spermatogenesis in male *G. aculeatus* took place when fish were exposed to the combination of short photoperiods and high temperatures (18°C), although the combination of short photoperiods and low temperatures characteristic of winter produced quiescent testes (Borg, 1982a). At the end of the breeding season, spermatogenesis restarts but the secondary sexual characteristics are lost. In male *G. aculeatus*, treatment at this time with androgen inhibited the onset of spermatogenesis although the secondary sexual characteristics were retained. This suggests that there is a negative correlation between spermatogenesis and the presence of high levels of androgens (Borg, 1981).

At no time in the annual cycle do the testes form much more than 1.5 per cent of the total body weight, and in *G. aculeatus* they reach their maximum relative size in autumn, that is, at the time of active spermatogenesis, not during the breeding season (Borg, 1982a), in sharp contrast with the female (Figure 7.1).

The secondary sexual characteristics of the male, which depend on the androgens produced by the interstitial cells of the testes, include the transformation of the kidney into an organ that produces the mucus used to glue the nest together, the breeding coloration and the organisation of reproductive behaviour. So, although by winter the male has sperm which are capable of fertilising the eggs, behaviourally the male is still sexually immature.

A symptom of the sexually maturing male is the increase in the height of the cells of collecting ducts and the distal portion of the proximal tubules of the kidneys. As the cells increase in height, they become transformed for the production of the glue used in nest building (Mourier, 1970). This transformation can be stimulated by treating females or castrated males with testosterone. It was even produced *in vitro* by treating cultures of kidney cells with androgen [11-ketotestosterone was particularly potent (De Ruiter and Mein, 1982)]. This transformation of the ducts and tubules is also marked by an increase in the relative size of the kidneys. Once the glue has been synthesised by the kidneys it is stored in the urinary bladder before use.

Males of *S. spinachia* and *A. quadracus* do not develop marked breeding colours, although the latter have prominent, red pelvic spines. The other species do, although within a species there is considerable inter- and intra-population variation in the form and degree of expression of the breeding colours. These male breeding colours can be induced in castrated males by treating them with testosterone. In *C. inconstans* and *P. pungitius*, the male darkens and in some populations becomes almost completely black with conspicuous white pelvic spines, but in other populations the extent of the darkening is much less. Foster (1977) compared the development of the male breeding colours in populations of *P. pungitius* from Europe (River Thames), the east coast of Canada and Lake Huron. During courtship, the European males were completely black, the coastal form had a grey back with a black ventral patch, and most of the Lake Huron males had saddle-shaped brown markings on the back, with black ventrally. At their fullest development, the breeding colours of male *G. aculeatus* are spectacular. The back becomes greenish-blue, the iris of the eye is blue and the anterior ventral surface is red. There is considerable variation in the extent of the development of each of these characteristics and in some populations the breeding colours may be absent, reduced or unusual (Chapter 10). Male *G. wheatlandi* adopt an overall greenish-gold breeding colour, usually with a series of black spots behind the pelvic spines (Wootton, 1976).

Although castrated males may be aggressive, they do not build nests or defend a territory around the nest, nor do they court females. Castrated males will show these behaviour patterns if they are treated with androgen. A reproductively competent male stickleback is one in which the testes are producing androgens. The interstitial cells are stimulated to produce these levels of androgen by gonadotrophin from the pituitary.

Reproductive Behaviour of the Sticklebacks

The reproductive cycle of the male sticklebacks can be divided into a number of behavioural phases. The initial phase is selection of a breeding site. This is followed by establishment of a territory. The nest-building phase then starts and when the nest is completed the male enters the sexual phase during which he is willing to court females. As a result of successful courtships the male then enters the parental phase. Upon completion of the parental phase, another cycle may be started. Behaviourally, the reproductive cycle of the female is much simpler. Most of her time is spent foraging. Once a batch of eggs has matured, courtship follows and can be completed quickly so that the female can resume feeding.

Selection of a Breeding Site

The selection of the breeding site is made by the males for it is they that settle to defend territories and build nests. Males often precede the females on to the spawning grounds (B. Borg, pers. comm.; Fitz-Gerald, 1983). Although laboratory tests have shown that male stickle-backs can show clear preferences for where they build their nests, observations on natural populations indicate that the males may be quite versatile in their selections, depending on the habitats available.

Selection in Gasterosteus. Both *G. aculeatus* and *G. wheatlandi* build their nests on the substrate, but the substrate used can vary from soft silt to hard rock. This ground-nesting habit frees *Gasterosteus* from a dependence on the presence of vegetation, but the ability to nest in the

open probably makes the male and his nest more vulnerable to predation. Laboratory experiments with *G. aculeatus* found that if sexually
mature males were released in a tank that had a sandy bottom, they
settled and started to build nests more quickly than if the tank had a
bare glass bottom or if the sand was covered with glass. More males
built nests on the sandy bottom than on glass (Schutz, 1980). Some
males will attempt to build nests on glass even though it is difficult
for them to attach nesting material to such a smooth substrate. In the
Little Campbell River (British Columbia), the distribution of breeding
males of the low and completely plated morphs of *G. aculeatus* suggested that the males chose different substrates on which to nest.
In laboratory experiments in which males were given a choice between
sand and mud as a substrate, the completely plated morph preferred
the sand but the low plated morph chose the mud (Hagen, 1967).
Unfortunately, no experiments were done to compare the reproductive
success of males that built on the different substrates.

Although *Gasterosteus* tends to select more open habitats in which
to breed than the other sticklebacks, there is both experimental and
field evidence which shows that the males will choose sheltered or
concealed positions in which to build their nests. These studies also
show that males are not restricted to such sites. Males of the low
plated morph of *G. aculeatus* collected from two lakes and a stream
in Washington State showed a clear preference for nesting in concealment in the form of vegetation or a tin box when tested singly in an
aquarium or in groups in concrete pools (Kynard, 1979a). Similarly,
males belonging to the completely plated morph collected from salt
marshes on Long Island preferred to nest inside clay flowerpots when
kept in small groups (Sargent and Gebler, 1980). In Lake Wapato,
in Washington State, nesting males tended to be most abundant in
habitats that provided some concealment, so that although there was
open habitat that was available for nesting it was underutilised by
the males (Kynard, 1978a). In the Little Campbell River, although
the nesting males of the low and completely plated morphs had different habitat preferences, they both tended to nest close to or in
stands of vegetation (Hagen, 1967). There is both experimental and
field evidence that males that build their nests in concealment are
reproductively more successful than males that build in the open
(Moodie, 1972a; Kynard, 1978a; Sargent and Gebler, 1980).

Gasterosteus wheatlandi may show a greater preference for cover
than *G. aculeatus*, for the former showed a significant preference for
nesting near vegetation in a large aquarium containing four evenly

distributed plants, whereas *G. aculeatus* showed no significant preference for nesting near plants planted in the same pattern in an aquarium
(Rowland, 1983a).

Males frequently have nests in shallow water, but males of the low
plated morph showed a clear preference for water deeper than 450 mm
when tested in groups in pools (Kynard, 1979a). Males in Lake Wapato
avoided nesting in water shallower than 150 mm and were seen nesting
in depths greater than 6 m. Nests that are built in very shallow water are
in danger of being uncovered if the water level falls (Wootton, 1971b)
or damaged by wave action during storms (Kynard, 1978a). Nevertheless, because sticklebacks are frequently found in shallow streams and
pools, the average depth of the nest is usually 150 mm to 500 mm
(Hagen, 1967; FitzGerald, 1983).

Selection in Pungitius. Pungitius is of particular interest. Typically,
the male builds his nest off the substrate in relatively thick vegetation.
However, in some populations in the Great Lakes, the nest is built on
the substrate. Males from populations in Lake Huron nest either under
or between rocks in shallow water (McKenzie and Keenleyside, 1970;
Foster, 1977), whereas males from a population in Lake Superior
nested in water several metres deep in a burrow dug in mud. The
ground-nesting *Gasterosteus* is absent from the Great Lakes, but *C.
inconstans* is present and this species also builds its nest in vegetation.

Selection in Culaea, Apeltes and Spinachia. Although these three have
very different forms of nests, they all select sites off the substrate
in vegetation. *Apeltes quadracus* from a tidal pool avoided nesting
on *Fucus* covered with filamentous algae, but were essentially general-
ists when selecting a nest site (Courtenay and Keenleyside, 1983).
All three species will build nests attached to air tubes in aquaria which
lack rooted vegetation (Figure 7.3) (Reisman and Cade, 1967; Rowland, 1974a; R.J. Wootton, personal observation).

Selection in a Multi-species Community. There are extensive areas
of sympatry between the various stickleback species (Chapter 2) so
at some sites breeding males of two or more species can be collected
from the same restricted area. The territorial behaviour of the males
makes it likely that there will be inter-specific competition for territories and suitable nest sites. In an experimental situation in which
G. aculeatus and *P. pungitius* collected from freshwater sites in Belgium
were competing for nest sites, *G. aculeatus* avoided nesting in dense

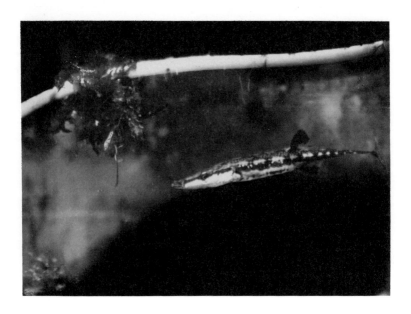

Figure 7.3: Male *S. spinachia* with His Nest Attached to an Air Pipe

vegetation but inhibited the nesting of *P. pungitius* in sparser vegetation (Ketele and Verheyen, 1982). Three species, *G. aculeatus*,
G. wheatlandi and *P. pungitius*, breed in salt-marsh pools in the St.
Lawrence Estuary. In one such pool in May, when there was little
algal growth, only the two *Gasterosteus* species were found nesting.
In June, algal growth was extensive and *P. pungitius* nested primarily within the algal growths whereas the *Gasterosteus* species were
usually in the more open areas (FitzGerald, 1983). In these natural
populations the species coexisted within the pools, but in an experimental study *G. aculeatus* was found to prevent *G. wheatlandi* from
establishing territories and could even displace resident *G. wheatlandi*
from territories they had already established (Rowland, 1983a). A
difference between the natural situation in the St. Lawrence Estuary
and the experimental situation was that, in the former, the proportion
of nesting *G. wheatlandi* was much higher than that of *G. aculeatus*,
but in the experiment equal numbers of the two species were present.

These inter- and intra-specific differences in the selection of breeding sites potentially provide considerable scope for the experimental
and observational analysis of the significance of such variation for
the reproductive success of the males.

Aggression and Territorial Defence

The male builds his nest within an area which he defends actively. Although this defence of his territory tends to be directed most strongly towards conspecifics, other species of sticklebacks and other intruders are also attacked. Territoriality has two components: attachment to an area, and aggressiveness, which defines how intensely that area is defended. These two components may vary independently so that the area defended may change, but the intensity with which a given area is defended may not change or may change in a different direction (Thresher, 1978). Aggression in sticklebacks is not restricted to that shown by reproductively mature males defending a territory.

Aggression in Juveniles

Juvenile *G. aculeatus* will attack other juveniles. Male and female juveniles did not differ significantly in the time they spent attacking a conspecific enclosed in a transparent tube, nor was juvenile aggression affected by whether the juveniles had been raised in isolation or in groups (Sevenster and Goyens, 1975). It did tend to be higher when the young were kept at a high density. There were also marked dissimilarities in the levels of juvenile aggression shown by fry from different parents, which suggests that there may be a genetic component to the differences in aggression observed between individuals (Goyens and Sevenster, 1976). In these experiments the juveniles belonged to the low plated morph from a population which was resident in fresh water throughout the year. In contrast, the juveniles from an anadromous population showed practically no aggression. This suggests that juvenile aggression may serve to disperse the young in freshwater populations, whereas in anadromous populations the young tend to migrate seawards in large schools (Goyens and Sevenster, 1976). In Heisholt Lake, British Columbia, underwater observations indicated that some sticklebacks maintained feeding territories outside the breeding season, which also suggests that, at least in some freshwater populations, aggression serves to disperse the population outside the breeding season (Maclean, 1980).

When the aggressive behaviour of males and females was tested in the month before they became sexually mature, the males showed significantly more lunges and bites at a stimulus fish than the females, but the latter threatened the stimulus fish more often. However, the organisation of the behaviour of males and females was relatively similar (Huntingford, 1979). At the time of sexual maturity, male fish that had been tested when they were juveniles showed significantly higher levels of aggression, whereas females showed significantly lower levels than they had as juveniles (Sevenster and Goyens, 1975).

Aggression in Mature Females

Although, in natural populations of *G. aculeatus*, females in the breeding season are usually in schools (Black and Wootton, 1970; FitzGerald, 1983), females kept in confined conditions will show aggressive behaviour (Wootton, 1976). When six females were kept in a 1200 × 1220 × 600mm tank, they developed dominance relationships, with the dominant females tending to have larger home ranges and being more aggressive. The dominant females also seemed to have preferential access to nesting males, and were more effective at disrupting the courtship of other females than the subordinate fish (Li and Owings, 1978a). This observation is significant because there is now evidence from natural populations that, particularly towards the end of the breeding season, gravid females may actively compete for the attentions of a nesting male (Kynard, 1978b; B. Borg, pers. comm.).

In laboratory tanks *P. pungitius* females show territorial behaviour (Morris, 1958), but such behaviour has not yet been recorded in natural populations.

Organisation of Aggressive Behaviour

In most aggressive interactions, the attacker approaches its victim rapidly and directly and attempts to bite it. Usually the victim flees. There are some situations, especially when two males are fighting close to the border of their territories, where the fighting may become more elaborate. The two fish may start to circle each other rapidly, often with their dorsal and pelvic spines erect (roundabout fighting). Threatening is also frequent. In *G. aculeatus*, the threatening fish adopts a head-down posture with its flank directed towards his opponent. The dorsal spines and the pelvic spine nearest the opponent are usually erect. Frequently, both fish will adopt this posture and move towards the substrate, jerking up and down rapidly. Sometimes a fish will dig into the substrate while in the threat posture. *Pungitius pungitius* has a similar threat display. *Culaea inconstans* and *Apeltes quadracus* will adopt a head-down posture while facing the opponent; in the latter species this orientation emphasises the conspicuous, erect, red, pelvic spines of the sexually mature male (Wootton, 1976).

The use of multivariate statistical techniques, especially principal components analysis, has suggested that it is meaningful to consider a group of behaviour patterns that together can be labelled 'aggression'. These patterns tend to be highly correlated with each other and probably share common causal factors (Huntingford, 1976a,b,c). The functional significance of these correlated behaviours is that a fish can use them to attempt to sequester resources for itself when in competition with other fish.

Stimuli for Aggression

In a classic series of studies, Tinbergen (1951) suggested that the aggression of the male *G. aculeatus* tended to be directed towards a stimulus which featured a red underside. Sexually mature males have a red throat and fore-belly. These early experiments used models of sticklebacks and the results suggested that as long as the model was not too large and the ventral surface was red, its shape did not have to be very fish-like. These results have been widely described in textbooks, but they are often misinterpreted as suggesting that aggression is only shown by males and only directed at stimuli with red bellies. Neither of these suggestions is true.

The aggressive behaviour of juveniles and females shows that aggression is not restricted to males, and males will attack other males, females, juveniles, other fish species and even invertebrates or other vertebrates. Anyone who has kept sticklebacks will know of males who have vigorously attacked a finger poked towards a nest containing eggs or fry. The level of the aggression tends to be higher when the intruder is a conspecific, especially when the male is still trying to acquire eggs for his nest, but the changes in the level of aggression during the breeding cycle tend to follow the same pattern irrespective of the nature of the intruder (Huntingford, 1977). These observations raise the question of the significance of the red coloration of males in aggressive encounters. Several studies have not confirmed that the aggression of males is greater when directed towards a stimulus that has a red belly (Muckensturm, 1967, 1968, 1978; Peeke *et al.*, 1969). In a more recent study, males attacked a model fish with a red belly less than a grey model (Rowland, 1982a). The effect of the redness of the stimulus is related to the context in which it is presented. In some circumstances, the red belly may intimidate a male and so result in a reduced level of aggression, whereas in other contexts, the red may augment the aggressive response of the male (Wootton, 1976; Rowland, 1982a). Tests in which males were paired in a small tank and a winner and loser of the subsequent fighting were noted showed that the most consistent predictor of which male won was the brightness of his colour

(Bakker and Sevenster, 1983). It is clear that a red belly does not act simply as a releasing mechanism for the male's aggression as originally suggested by Tinbergen, but it is equally clear that the red belly is an important cue that can be used by a male in making a response to an intruder. The red belly is certainly used as a cue by a female when courting a male.

Territoriality

Territorial behaviour evolves when a spatially localised resource can be defended economically (Davies, 1978). For the male stickleback, the nest as a spawning locality forms the defendable resource. Territorial defence involves both costs and benefits where both are measured in terms of fitness. In the case of the stickleback, the benefits are obvious: only territorial males leave progeny. The costs are less obvious but probably include an increased risk of predation, an increased risk of physical injury during territorial fighting and decreased access to food. A simple graphical model can illustrate the relationship between benefits, costs and territory size (Figure 7.4). Such a model suggests that males should only take up a territory if its size is sufficient to yield a net gain in fitness, that is, the benefits should exceed the costs. The model further suggests that there may be a minimum size below which there is no net gain, and also a maximum territory size above which, again, there is no net gain in fitness.

Both experimental and field studies on *G. aculeatus* have suggested that there is a minimum size of territory which males will accept. Solitary males kept in tanks smaller than 300 × 300 mm were significantly less likely to build nests, or, if they built nests, to complete courtships successfully than males in larger tanks. When 30 males were introduced into a 1500 × 1000 mm tank, the minimum territory size was about 450 × 450 mm so not all the males were able to establish territories (van den Assem, 1967). In natural populations, if there is relatively little cover so that neighbouring males can easily see each other, the minimum distance between the nests of neighbours is usually greater than 300 mm (Black and Wootton, 1970; Wootton, 1972; Kynard, 1978a). Smaller inter-nest distances occur where there is thick vegetation or where the topography of the bottom is such that neighbours are less visible. In an area of Lake Wapato, the mean inter-nest distance was 411 mm and the smallest distance was 150 mm where males were nesting in thick vegetation (Kynard, 1978a). In a salt-marsh pool in the St. Lawrence Estuary, the mean inter-nest distance was 332 mm for *G. aculeatus*, 545 mm for *G. wheatlandi* and 517 mm for

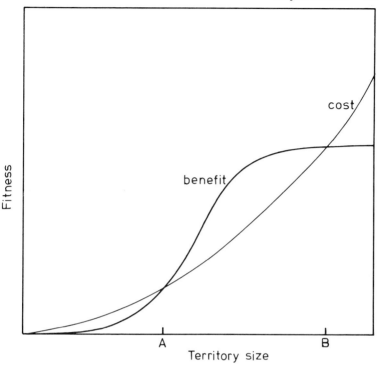

Figure 7.4: Schematic Diagram of Relationship between Fitness Costs and Benefits and Territory Size. A, Minimum Territory Size, and B, Maximum Territory Size at Which Costs = Benefits. After Stanley (1983)

P. pungitius (FitzGerald, 1983). In all these cases the distribution of nests was more regular than would be predicted if the nests were distributed at random with respect to each other. The presence of a male with a nest significantly reduces the chances of another male building a nest in the close vicinity. There is no equivalent information on the maximum territory size beyond which the costs exceed the benefits. When pairs of male *G. aculeatus* were introduced simultaneously into a tank 6000mm long, the mean inter-nest distance was 71 per cent of the maximum possible distance (van den Assem, 1967).

The nest acts as a focus for the aggression of the male, so that a male *G. aculeatus* maintains a relatively high level of aggression to an intruder up to about 300mm from the nest, but beyond that the level of aggression declines with distance from the nest (van Iersel, 1958; Symons, 1965).

One mechanism that could reduce the cost to a male of maintaining a territory would be to minimise the aggression shown to males that have adjacent territories and so have established themselves as neighbours. Such a reduction in the level of aggression is shown by adjacent male *G. aculeatus*. After an initial period of high levels of aggression while the territories are being established, there is a subsequent decline in the aggression between neighbours. If a strange male intrudes into a territory, then it still attracts high levels of aggression (van den Assem and van der Molen, 1969). Male aggression does wane if the stimulus eliciting the aggression, be it a live fish or a model, is repeatedly presented to a male. This waning is specific to the particular stimulus and will occur even if the male is unable to actually show an aggressive response to the intruder (Peeke and Veno, 1973, 1976). Interestingly, this habituation of the aggressive response disappears if the male rebuilds his nest. It is as though the male 'forgets' that he has had experience of the stimulus that elicits his aggression if he has rebuilt his nest in the interval between successive appearances of the stimulus (Peeke *et al.*, 1979). Although there may seem to be some advantage in this 'dear enemy' phenomenon (the male is able to reduce the amount of aggression required to maintain his territory), in some situations neighbouring males are dangerous rivals, so that while they may be ignored if they are at the border of their own territory, they will be attacked vigorously if detected well inside the male's own territory.

Physiological Control of Aggression and Territoriality

Castrated male *G. aculeatus* are aggressive if kept under a long photoperiod (16L-8D), but the full expression of male aggression and territoriality does depend on androgens produced by the testes. Castrated males do not build nests, so their aggression lacks the spatial focus and intensity that is associated with the presence of a nest in intact males (Baggerman, 1968).

Nest Building

Although in all the stickleback species the repertoire of nest-building behaviours is similar, with the male using a mucus string produced by the kidneys to glue his nest together, the final form of the nest varies considerably between species. *Spinachia spinachia*, *Apeltes quadracus* and *Culaea inconstans* build their nests off the substrate in vegetation. *Gasterosteus aculeatus* and *G. wheatlandi* build nests on the substrate and associated with this ground-nesting habit have a richer repertoire of nest-building behaviours because it includes manipulation of the substrate. *Pungitius pungitius* includes both vegetation and substrate nesters (Wootton, 1976; Keenleyside, 1979).

Nest Building of Spinachia spinachia. This species builds the largest nest, the diameter typically being 50-80mm. The male selects a patch of seaweed, then lays down a network of thick threads of glue, spending as long as 15 min glueing. He then brings pieces of seaweed and pushes them into the network. Further long bouts of glueing follow at intervals of 1 to 5 h and material is added to the nest, which the male consolidates by pushing at it with his snout. The completed ball-like nest lacks a distinct entrance or exit tunnel (Figure 7.3).

Nest Building in Apeltes quadracus. In contrast, this species builds the smallest nest, which may be only 5mm in diameter. However, the species is unique in that nests are built one on top of the other forming a 'high-rise' suite of nests. Having selected a suitable site in vegetation, the male glues and brings material to the site. He produces a cup-shaped nest with only a few strands of material forming the roof. There is no distinct entrance tunnel. Once a female has spawned in this first nest, a second nest may be constructed above the first, and this second nest will receive the eggs from the next spawning. This process may be repeated so that a tier of nests is built, each containing one clutch of eggs. The male may even start a new nest at another site, while there are still eggs developing in the first nest-complex. In all the other species, the successive clutches are laid in the same nest, although this may be enlarged to accommodate them.

Nest Building in Culaea inconstans. The nest is neat, compact and small, 15-30mm in diameter. It is frequently placed in the fork between branches or attached to the stem of a plant. *Culaea inconstans* shares with *P. pungitius* two forms of glueing behaviour. In superficial glueing, a string of glue is secreted on to the outer surface of the nest. Superficial glueing is shown by all the species. When insertion glueing, the male *C. inconstans* catches a strand or drop of glue in its mouth, then places this glue on the inner surface of the nest. Insertion glueing is shown at the end of the nest-building phase and it probably consolidates the nest ready for the entry of the female. The male forms a cavity by pushing his snout into the next and then twisting it from side to side, but he does not make an exit tunnel.

Nest Building in Pungitius pungitius. The typical nest is built among the branches and leaves of vegetation. It is larger and more untidy than the nest of *C. inconstans*, but the process of building the nest is essentially the same. Insertion glueing appears towards the end of the nest-building phase, the end of which is marked by the male creeping through the nest forming a tunnel with a distinct entrance and exit. This creeping through is repeated at intervals until the male becomes parental. Males from a population in Lake Huron living on a rocky, wave-swept shore built almost token nests of algal and detrital fragments under or between rocks. Males from a population in Lake Superior built their nests in mud, constructing a burrow 30-40 mm long and 10 mm in diameter that had only a single entrance. These inter-population differences in nest building potentially provide scope for a genetic analysis of such behaviour.

Nest Building in Gasterosteus. The nest is invariably built on the substrate, usually on sand or mud, but not exclusively so. Initially, the male digs a pit in the substrate and then starts to bring material to this pit. The material is glued together by superficial glueing. By pushing at the nest, the male shapes it, and he may anchor it to the substrate by digging small pits around the periphery. He bores at the front of the nest to form the entrance to a tunnel and the end of nest building is marked by the male creeping through the nest to form a complete tunnel. When given a choice of material, the males tends to prefer coarse, thin, loose threads with a length of 20-40 mm (Schutz, 1980), but males are versatile at adapting many different types of material to their purpose. A male can successfully build a nest on a hard substrate where a nest pit cannot be excavated and will use any scraps of vegetation and detritus that are available. Even if the substrate is covered with a plate of smooth glass and no vegetation is present, some males will attempt to build a nest, glueing and bringing detritus to the site (Wootton, 1976; Schutz, 1980).

Behaviour in the Sexual Phase

The sexual phase is the period in which the male actively courts gravid females and acquires clutches of eggs in his nest. Although the basic organisation of courtship behaviour is similar in all the sticklebacks, there are considerable inter-specific variations in the details. The courtship of *G. aculeatus* has been studied in most detail, but descriptions are available for the other species, although for some the information is sparse (Wootton, 1976; Keenleyside, 1979).

Courtship probably has several functions. It advertises the presence of a sexually receptive fish. It also provides an opportunity for mate choice and so reduces the chances that gametes will be wasted. This function is particularly important in maintaining reproductive isolation between closely related and morphologically similar species such as the sticklebacks. The courtship results in a spatial orientation of the male and female so that the eggs and sperm are deposited at the correct site. The activities of the male and female are synchronised ensuring that the release of eggs and sperm occur within a short interval of each other. Courtship in sticklebacks may also play a role in reconciling a motivational conflict which may develop because the female represents a potentially cannibalistic intruder into the male's territory, as well as being a possible sexual partner. This aspect of courtship seems to be particularly important in *G. aculeatus*.

Courtship in Spinachia spinachia. This is the least well known species. The few descriptions of the courtship suggest that the male approaches a female which has entered his territory and nudges her on the tail stalk or fins. If the female does not flee, the male swims directly to his nest and shows it to the female by placing his snout in the nest and then jerking his snout from side to side. If the female is ready to spawn, she follows the male and pushes her way into the nest. The male then takes up a position at right angles to her and gently bites her tail stalk until she spawns. After the female has spawned, she leaves the nest and the male pushes through, fertilising the eggs.

Courtship in Apeltes quadracus. When a gravid female enters the male's territory, he approaches erecting his prominent red pelvic spines. Although his initial approach is slow, he then leaps forward and pummels the female with his snout. The female sinks to the bottom but adopts a head-up posture as the male swims rapidly away following a horizontal, spiral path. He stops between the female and his nest and adopts a head-down position displaying his pelvic spines while facing his nest. If the female does not swim to the male, further bouts of pummelling and spiralling occur, and the male may return to the nest and perform some nest-orientated activities. When the female does swim to the male, he swims slowly towards the nest in short jerks. The female follows closely behind and underneath the male, perhaps visually fixating on the pelvic spines. At the nest, he presses his snout into the nest entrance and gives a few strong tail beats. The female pushes her way in, thrusting diagonally upwards with the nest forming a flimsy ring round her belly. The male then gently takes her tail fin

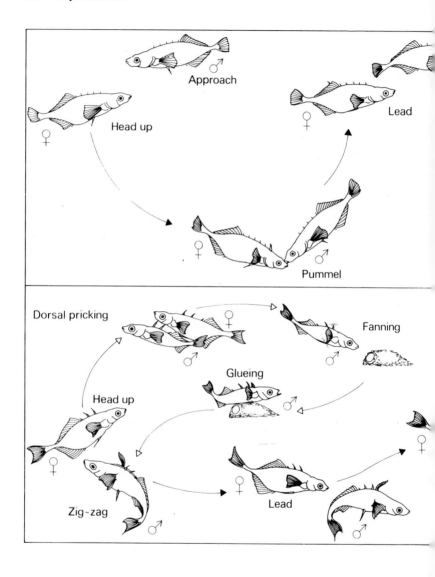

Figure 7.5: Courtship in the Vegetation-nesting *C. inconstans* (Top Sequence) and the Substrate-nesting *G. aculeatus* (Bottom Sequence). Closed Arrows, Approach-Lead Sequence; Open Arrows, Dorsal Pricking Sequence. After Wootton (1976)

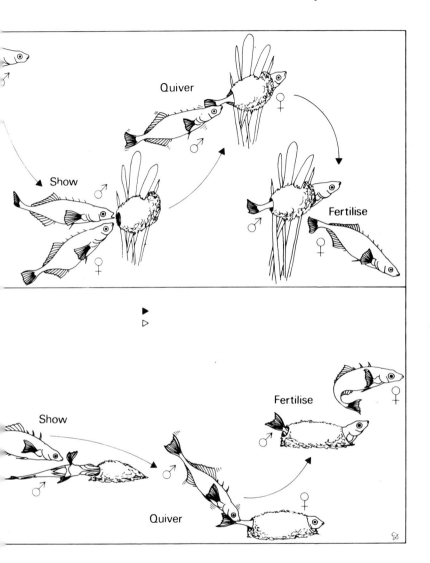

or stalk in his mouth and quivers. This quivering is repeated until the female spawns. Then the female squirms out of the nest and the male creeps through it, fertilising the clutch. The female is chased away and the male starts to build the next tier of the nest.

Courtship in Culaea inconstans. The courtship of this species also starts with the male pummelling the female towards the bottom where she assumes a head-up posture. The male swims towards his nest with his spines erect and back arched. He moves slowly with exaggerated beats of the tail fin. If the female does not follow, he returns to pummel her again. When the female does follow, she stays just below and behind him. At the nest he places his open mouth against the upper rim of the nest and fans by rapidly beating his tail fin but preventing forward movement with compensatory beats of the pectoral fins. (This fanning is typically used by male sticklebacks to drive a current of water through the nest to ventilate the eggs.) The female rides the current created by the fanning then pushes her way into the nest. The male places his snout against the flank of the female and quivers rapidly. Bouts of quivering follow until the female spawns. After spawning, the female forces her way out of the nest through the back wall and the male creeps through, fertilising. Immediately after the spawning, the male shows high levels of fanning, but then he repairs and enlarges the nest so that it is ready to accommodate the next clutch of eggs (Figure 7.5).

Courtship in Pungitius pungitius. When the male notices a gravid female in his territory, he adopts a head-down posture and dances towards her in a series of jumps or zig-zags. Each jump has a forwards, an upwards and a sideways component. With the approach of the male, the female assumes a head-up posture which makes her distended belly prominent and moves towards the male. He dances back to the nest with the female following slightly below him. Sometimes, the male does not dance directly back to the nest but follows a circuitous path. At the nest, he assumes a head-down position with his snout just above the nest entrance and fans. Once the female has entered the nest, he quivers in the same way as *C. inconstans* at this stage of the courtship. After spawning the female swims out through the exit of the nest tunnel and the male fertilises the eggs and then drives the spawned-out female away. An interesting variation is shown by the ground-nesting Lake Huron *P. pungitius*. Although most of the courtship is similar, the male tends to lead the female directly back to the nest rather than dancing back. The Lake Huron males also perform fewer zig-zags in their approach to the female (Foster, 1977).

Courtship in Gasterosteus wheatlandi. The male approaches a gravid female in a series of zig-zag jumps executed with the male in a head-down posture. The female adopts the receptive head-up posture and orientates to the male. He leads her back to the nest by maintaining his head-down posture while holding his body so that his back is arched. As he swims slowly to the nest he quivers continually. The female maintains her head-up posture but follows the male from underneath from where she can nudge the male between his pelvic spines. The male does not take a direct route back to the nest and by adopting a devious pathway the male can make the leading phase last as long as a minute. At the nest, the male shows the nest entrance and induces the female to spawn by quivering on her flanks.

Courtship in Gasterosteus aculeatus. This has been made one of the famous courtships in biology by the classic, ethological studies of Tinbergen and his students. Many textbooks use the courtship of this species as an example of a reaction chain in a sequence of behaviour in which the action of one of the partners acts as the stimulus that evokes the next response of the other partner and so on until the behavioural sequence has been completed. In reality, the courtship of *G. aculeatus* is more complicated and variable than the original reaction-chain model suggests.

At its simplest, the courtship sequence can closely follow the reaction-chain model. When the male sees a gravid female in his territory, he approaches her with a zig-zag dance, in which the male first jumps away from and then towards the female. In contrast to *G. wheatlandi* and *P. pungitius*, the jumps are not performed in a head-down posture. The female adopts the receptive head-up posture and swings towards the approaching male. He turns and swims directly back to the nest and the female follows above and slightly behind him. At the nest, the male pokes his snout into the entrance and rolls on to his side so that his back is towards the female. He jerks to and fro, 'showing' the entrance. She pushes past him into the nest and he quivers on her flanks until she spawns. She leaves the nest and the male creeps through, fertilising the eggs (Figure 7.5). This sequence can break down at any stage, or steps may be repeated or omitted.

There is, however, a major deviation from this sequence. Instead of leading the female back to the nest, the male frequently shows a behaviour called dorsal pricking. As the female, in her head-up posture, tries to follow the male, he positions himself below and across the female and with his dorsal spines erect moves backwards as if he is trying to prick the distended belly of the female with his dorsal spines. The two fish usually circle while the male shows this dorsal pricking. The male then breaks away and swims back to his nest, but in this case, if the female attempts to follow, he either resumes dorsal pricking or starts attacking her. Usually the female waits in her head-up posture when the male swims down to his nest. Here, he performs nest-orientated activities. Usually, early in the courtship he creeps through the nest; subsequently, he is likely to bore into the nest entrance, then fan and finally glue. After creeping through, but especially after a sequence of bore-fan-glue, the male will zig-zag vigorously towards the female and attempt to lead her back to the nest (Figure 7.5). The curious thing about this dorsal-pricking, nest-visit sequence is that contact between the male and female is broken off and the female is left on the periphery of the territory while the male returns alone to his nest. There are no descriptions of behaviour patterns similar to dorsal pricking for the other species, although in other respects the general organisation of the courtship sequence is similar. Under what circumstances does this form of courtship behaviour occur and what is its functional significance in *G. aculeatus*?

At this stage in his reproductive cycle, the male has to reconcile several different behavioural requirements. He must be ready to court and lead to his nest any gravid female that enters his territory, and to repel any intruder from it, even though the intruder and a gravid female will represent relatively similar stimulus patterns. He must also maintain his nest in a suitable condition for it to receive eggs if his courtship is successful. Dorsal pricking and nest visits can be seen as a behavioural mechanism that enables these requirements to be reconciled.

If the male is made relatively aggressive by presenting him with another male as an intruder just before he is presented with a gravid female or a model of a female, he tends to show high levels of dorsal pricking and nest visits before he switches to attempting to lead the female. If the male is exposed to another female before being tested with a gravid female, he shows little or no dorsal pricking, but high levels of zig-zagging and leading. This suggests that the nest visits, the creeping through and the bore-fan-glue sequence enable the male to switch from a primarily aggressive motivation to a primarily sexual motivation. It is noticeable that the male changes colour after creeping through, tending to become much paler, which might be interpreted as a symptom of a change in his motivational state. The dorsal pricking serves to inhibit the female from following the male back to the nest and allows him to perform the nest activities which result in the motivational shift required for a successful courtship (Wilz, 1970a,b,c). A comparison of a population from the east coast of North America with one from England suggested that the males from the American population showed much more dorsal pricking in their courtship than the English males. There was also evidence that the American males tended to be more aggressive in other situations than the English, which is compatible with the hypothesis that dorsal pricking occurs when the male is relatively aggressive (Wilz, 1973).

Other species do not seem to show the equivalent of dorsal pricking in their courtship; indeed the pummelling component in the courtship of *A. quadracus* and *C. inconstans* suggests that they are relatively aggressive during a courtship. A comparison of *P. pungitius* and *G. aculeatus* suggested that the general level of aggression is much lower in the former, so the problem of switching from an aggressive to a sexual motivation is less acute in *P. pungitius*. This species also tends to occur in heavily vegetated areas where it is likely that the separation of male and female during the courtship would lead to the partners losing contact with each other (Wilz, 1971). Experimental studies have suggested that *G. aculeatus* is also more aggressive than *G. wheatlandi* or *A. quadracus* (Rowland, 1983a,b).

In *G. aculeatus*, the frequency of nest visits tends to increase during a courtship. The other species may also return to the nest without the female following and may show some nest-directed activities, but these nest visits tend not to increase in frequency during a courtship. Such visits may help the male to keep the nest in a suitable condition to receive the female and may enable him to detect the presence of any intruders attempting to interfere with his nest (see below). The importance of these nest visits was suggested in experiments that used a tank divided into three compartments linked by doors. A male *G. aculeatus* was allowed to build a nest in one end-compartment, and this male could court a gravid female held in a container in the other end-compartment. The male could be trapped for a controlled period of time in the middle compartment when on his way to the nest or on his way to the female. The experiments suggested that the male was sharing the available time between female-directed activities and nest-directed activities. When the nest was disturbed by placing snails on top of it, males whose dominant activity was nest-directed showed a significant decrease in the rate of zig-zagging and in the time spent in normal nest activities. Males whose dominant activity was female-directed showed no significant decline in zig-zagging or in the time spent performing nest activities. Further analysis suggested that during courtship, three behavioural systems competed for dominance, aggression, sex and nest activities (Cohen and McFarland, 1979). Comparable experiments have not yet been published on the other species in which nest activities tend to decline during courtship, although the problem of reconciling the three behavioural systems will still be present.

Creeping-through cycle in Gasterosteus aculeatus. The first occurrence of creeping through (CT) marks the transition from the nest-building to the sexual phase in the male. But CT recurs at intervals during the sexual phase and is frequently shown during courtship, particularly in the first minute or so. If a gravid female is presented to a male for a short time but removed once the male has crept through, successive CTs occur, with the time interval between each increasing geometrically. If the female is not removed but is prevented from spawning by confinement in a glass tube, the interval between successive CTs remains approximately constant. A simple model can account for these observations. Assume two variables, E and T. E (for excitation) is increased by the presence of the female, and T (for threshold) is increased by the act of CT. Assume that in the absence of a female, E decays exponentially, and that after CT, T also decays exponentially but at a faster rate than E. CT occurs when $T = E$. If the female is constantly present, E remains relatively constant, but T is raised to its maximum levels after each *CT* and decays exponentially. This model accurately predicts the temporal pattern of CTs, including the length of the interval between two successive CTs (ICTI) (Nelson, 1965; 't Hart, 1978). During each ICTI, the behaviour of a male changes in a predictable way. Immediately after CT, the male has a high tendency to zig-zag, even towards inappropriate stimuli. This tendency declines with the approach of the next CT. Fanning at the nest is very low after CT but increases, tending to reach its highest levels immediately before the next CT. Aggression is also low immediately after CT but tends to increase during the ICTI (Sevenster, 1961; Nelson, 1965; Wilz, 1975). The striking feature of this creeping-through cycle is the regularity of the events within it.

The length of the ICTI can be shortened by reducing the length of the nest, but lengthening the nest does not increase the length of the ICTI. This suggests that during a CT the fish receives information that sets the length of the ICTI, but there is a maximum length which cannot be exceeded. In terms of the model, the height of T is set by the time spent creeping through, but the nest is normally built so that a CT raises T to its maximum level ('t Hart, 1978). Although the CT cycle has attracted much ingenious experimental analysis, its function during the sexual phase of the male is obscure. Why should the events of the cycle be so regular? What is the significance of the changes in the frequencies and durations of zig-zagging, fanning and other nest activities in the intervals between CTs? A creeping-through cycle or its equivalent has not yet been described for the other species, even though in *G. wheatlandi* and *P. pungitius* the male does creep through the nest to form a complete tunnel.

Mate Choice during the Sexual Phase

During courtship, the male and female have the opportunity of choosing a mate, and the choice will be an important factor in determining the reproductive success of the partners. The role of mate choice in restricting gene flow between plate morphs and in other polymorphic situations will be discussed in Chapter 10, but even within a population, there is scope for mate choice. When presented simultaneously with two models of gravid females, *G. aculeatus* males directed more courtship at the model with the more distended abdomen. If the models were presented consecutively, there was no difference in the amount of courtship directed at the two models (Rowland, 1982b). A female with a highly distended belly will probably spawn more eggs than a female that is less swollen with eggs. Especially towards the end of a breeding season, several females may attempt to court nesting males simultaneously (Kynard, 1978b; B. Borg, pers. comm.), so the male is in a position to make a choice.

Female *G. aculeatus* showed a preference for mating with a male with a red throat compared with sexually active males in which the red throat was not developed (Semler, 1971).

Mate choice during courtship is also important in maintaining the reproductive isolation of stickleback species when they are sympatric. Sexually mature *G. aculeatus* and *P. pungitius* are frequently found in close proximity and gravid females of the two species are morphologically similar. Males of the two species were found not to discriminate between females of the two species in terms of the rate of zig-zagging. In most cases, the females ignored the courtship of the heterospecific male, although they were receptive to a conspecific. Female *P. pungitius* were slightly less choosy than female *G. aculeatus*, but even if they responded positively to an approach by a male *G. aculeatus*, the response took a long time to develop and the courtship tended to break down because of the aggressiveness of the male or during the lead to the nest. Two *P. pungitius* females did spawn in male *G. aculeatus* nests but few eggs hatched and the fry all died within 2 days. The females could use as cues to discriminate between males of the two species early in a courtship both the breeding colours and the orientation of the male during the zig-zag dance. It was not clear which of the two was more important (Wilz, 1970d, 1971). In natural situations in which two or more stickleback species are sympatric, there does not seem to be any hybridisation between species even when there is considerable spatial and temporal overlap during breeding. This suggests that the distinctive courtships of the species act as effective reproductive isolating mechanisms.

Behaviour during the Parental Phase

Once the male stickleback has been successful in courtship and has eggs in his nest, he assumes a parental role (Wootton, 1976; Keenleyside, 1979). For a parental male the dominant activity is ventilating the eggs. In the absence of some form of ventilation, the water inside a nest is likely to become stagnant and deoxygenated as the eggs respire. Eggs may also get covered in silt and debris. Except in unusual circumstances, stickleback eggs in a nest will not develop to hatching without the male ventilating them. In all species except *A. quadracus*, the male ventilates his nest by fanning, which is similar in the species that show it. The tail and hind part of the body move as though to drive the fish forward, but the broad pectoral fins counter this forward-thrust and in doing so push water forwards over and through the nest. Fanning is such a distinctive activity that it can be recognised even when it occurs, as it occasionally does, away from the nest. It first appears in the behavioural repertoire towards the end of the nest-building phase and also occurs during the sexual phase, but only for relatively short periods of time. The main inter-specific difference is that the ground-nesting *Gasterosteus* species fan in a head-down position, whereas the species that nest off the substrate fan in a horizontal position (Figure 7.6).

Apeltes quadracus ventilates its nest complex in a quite different way: it sucks. The male places his snout in the nest entrance and by gulping draws water into his open mouth and expels it through the opercular slits, so the water moves in an opposite direction through the nest opposite to that caused by fanning. Fanning is a vigorous activity and *A. quadracus* has the smallest and most delicate form of nest, so the evolution of the small nest may have required the evolution of a less powerful method of ventilating it.

During the parental period, the male shows other forms of parental care. He eats eggs that have died and have become mouldy. As the eggs become metabolically active, the nest is altered in a way that probably eases their ventilation. In *G. aculeatus*, the male makes holes in the roof and around the rim of the nest. When the young hatch, the nest is torn apart so the newly hatched fry lie in a tangled mass of vegetation.

Males of *C. inconstans* and the plant-nesting *P. pungitius* construct 'nurseries' of loosely entwined pieces of vegetation either out of or above the nest. As the young fry become free-swimming, the male picks them up in his mouth and spits them into the nursery. *Gasterosteus aculeatus* will also retrieve young fry and spit them back into the nest pit (Feuth-de Bruijn and Sevenster, 1983).

Temporal Organisation of the Parental Phase. The length of the parental phase depends on the rate at which the eggs develop, which is a function of temperature. At 10°C, it takes 15-20 days from fertilisation for the eggs of *G. aculeatus* to hatch, but at 20°C, hatching is in 6-8 days. The length of time for which the fry are guarded is variable. In Lake Wapato, the number of days the male stayed with the fry after hatching varied from 1 to 9. The average length of a male's reproductive cycle was 13-14 days when the surface water temperature was 21°C. One or two days were spent in the nest-building and sexual phases, it took 5-6 days for the eggs to hatch, and about 6 days were spent caring for the fry (Kynard, 1978a).

In the period between the fertilisation of the eggs and the fry dispersing, the male's behaviour changes in well defined patterns. These patterns have been most completely described for *G. aculeatus* but similar changes are shown by the other species though with some modifications.

Immediately after fertilising a clutch of eggs, the male *G. aculeatus* manipulates them so that they become a flattened mass and another clutch can be laid partially overlapping them. The nest is repaired and extended forwards. During this period, the male is aggressive and shows little or no courtship to other gravid females. After about an hour, the tendency to court females has recovered and the male attempts to acquire more clutches. Over a period of 2 or 3 days, the male may obtain several clutches. In Lake Wapato, males in May had an average of about 1000 eggs in the nests, and one had 2680! At this

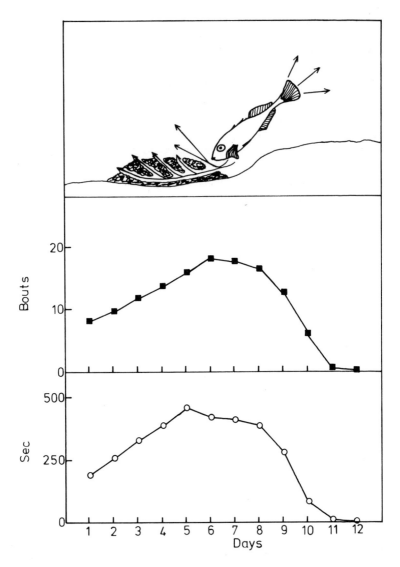

Figure 7.6: Fanning of Male *G. aculeatus* during Parental Cycle. Top, Schematic Diagram of Orientation of Male and Direction of Water Currents; Middle, Mean Number of Fanning Bouts per 15 min; Bottom, Mean Seconds Spent Fanning Per 15 min. Means Based on 13 Males, Each Fanning a Single Clutch of Eggs

time, the mean clutch size of a female was 96 eggs (Kynard, 1978a). Males in Trout Lake in British Columbia had an average of six clutches in their nests, whereas males from Garden Bay Lake in the same area had only three (Pressley, 1981).

Once the male has eggs in his nest, the amount of time spent ventilating the nest increases and reaches a peak just before the eggs hatch (Figure 7.6). Both the number of bouts of fanning and the mean length of each bout increases. After hatching, there is a rapid decline in the time spent ventilating. A solitary male *G. aculeatus* can spend over half of his time fanning just before the eggs hatch. Comparable temporal patterns are shown by *G. wheatlandi*, *P. pungitius* and *C. inconstans* but these species do not reach such high peaks of fanning. A similar pattern is seen in *S. spinachia* but the parental cycle is longer so that the changes are stretched out over a longer period of time. Although the basic pattern is also similar in *A. quadracus*, because of the complex structure of the male's nest, different nests contain eggs at different stages of development. The male spends most time ventilating the nests that have eggs in the most advanced state of development, but he will switch from one nest to another to provide some ventilation for all the clutches he has collected. There are no reports of the male *A. quadracus* showing any form of retrieving behaviour once the fry have escaped from a nest.

Experimental studies with *G. aculeatus* show that the changes in the amount of fanning depend both upon the presence of eggs in the nest and the number of fertilisations the male has experienced. In the early days of the fanning cycle, the male can be stimulated to fan more by running water through the nest that is relatively rich in carbon dioxide and relatively poor in oxygen. The amount of fanning is also increased the more clutches there are in the nest and the more fertilisations the male has experienced. As the fanning cycle progresses, its pattern comes to depend rather more on changes that are internal to the male and less on the stimuli the male is receiving from the eggs. The male becomes established in a parental mode of behaviour. If the male loses the eggs in the nest early in a parental cycle, he will switch back into a sexual phase and resume courting females. If the eggs are lost later in a parental cycle, the male does not resume courting but destroys the nest and builds a new one (van Iersel, 1953; Wootton, 1976).

The establishment of the parental mode in *G. aculeatus* is also shown by the appearance of retrieving behaviour. Experiments in which day-old fry were placed close to the nest of a male showed that the tendency to retrieve young develops some days before the male's own eggs have hatched. If the male was hungry, there was a tendency for retrieving to develop later in his parental cycle. Early in the cycle he ate the strange young, but even hungry fish retrieved at about the time their own young were hatching. When perch fry were used in the test, they were eaten immediately even by males with their own fry in their nests. The fry of *P. pungitius* were retrieved by satiated males but only with hesitation and after apparently testing the fry in their mouths. Hungry males ate the *P. pungitius* fry but still retrieved *G. aculeatus* fry around the time that their own eggs hatched (Feuth-de Bruijn and Sevenster, 1983).

Courtship behaviour declines as the males move into the parental phase. This decline is more rapid the more clutches the male has and the more fertilisations he has experienced. Males that had experienced five fertilisations and retained five clutches showed little or no sexual behaviour 4 or 5 days after the first fertilisation (van Iersel, 1953). In the process of becoming parental, the male sacrifices the possibility of fertilising more clutches. In *A. quadracus*, the decline in sexual responsiveness seems to be less developed and the male may retain the ability to court females and fertilise their eggs for 2 or 3 weeks after the first fertilisation.

Throughout the parental phase, the male continues to defend his territory aggressively. The nest with eggs or fry is an attractive focus for both conspecific and heterospecific predators. Because of the time the male must spend in ventilating and other parental activities, the time available for patrolling his territory and attacking intruders is much less than in the sexual phase. At least in *G. aculeatus*, the male solves this problem by maintaining high levels of aggression close to the nest, while essentially abandoning the outer portions of his territory. When presented with an intruder male at various distances from the nest, a parental male shows a high and relatively constant level of aggression towards the intruder at distances up to about 300 mm from his nest. At greater distances, the aggression tends to decline from the day on which the eggs were fertilised until about the day of hatching, but then the level of aggression increases significantly when the fry appear (Symons, 1965; Black, 1971; Wootton, 1971a). Territory size and the average inter-nest distance also tend to decline as the fanning cycle proceeds (Black, 1971; Wootton, 1972; Kynard, 1978a), a decline that is associated with the increase in the time the male spends at his nest as the eggs develop.

An increase in the aggression shown to heterospecific intruders also occurs around the time the eggs hatch. When the aggression of male *G. aculeatus* was measured before they had eggs in the nest, early in the parental cycle and when they had fry, there was a significant increase in the aggression shown in the parental period whether the intruder was a male *G. aculeatus*, a male *P. pungitius* or a female newt. The highest levels were shown when the eggs had hatched. At all times the level of aggression shown to the fish intruders was higher than that shown to the newt, but whereas initially the aggression shown to the conspecific male was higher than that shown to the heterospecific male, this differential response lessened as the parental cycle progressed (Huntingford, 1977).

This increase in the aggression of the male associated with the appearance of the fry was correlated with an increased 'boldness' towards a fish predator, the pike, *Esox lucius* (Huntingford, 1976c). In Lake Wapato, males that had eggs or fry in their nest were less likely to flee when a predatory rainbow trout (*Salmo gairdneri*) was manoeuvred near their nest than males with an empty nest. Some males caring for fry even showed aggressive displays towards the approaching trout (Kynard, 1978a). The spiny sculpin, *Cottus asper*, is a predator of both the eggs and free-swimming sticklebacks in some lakes on the west coast of North America. When a model sculpin was placed over the nests of male sticklebacks, the males that remained in their territory and attacked the model had a larger average number of eggs in their nests, or the eggs were, on average, older than those of the males that deserted their territory and never attacked the model. In a lake in which the sculpin occurred naturally, the males that attacked the head of the model — that is, the dangerous end — had older or more numerous eggs than males that attacked the tail. Both the time taken for a male to return to his nest and the time before the male started attacking the model decreased as either egg number or egg age increased (Pressley, 1981).

The changes in aggression and response to predators during the parental period, which have been seen both in laboratory experiments and in natural populations (especially the increase in aggression and boldness associated with the appearance of fry), indicate that the male takes higher risks as the contents of his nest become more valuable in terms of their contribution to the male's reproductive success. Early in the parental phase, a male that has lost his eggs simply reverts to the sexual phase and attempts to gather more, but later in the parental cycle, the loss of eggs means that the male must restart the whole reproductive cycle from re-establishing his territory and building another nest.

Physiological Control of the Parental Phase

Although there are dramatic changes in the behaviour of the male during his parental phase, the physiological mechanisms responsible for the control and integration of these changes to ensure a functional parental cycle are not known. Lesions in the telencephalon of the brain of male *G. aculeatus* caused changes in the temporal patterning of fanning, aggression and courtship during the parental cycle, which suggests that the telencephalon is implicated in the integration of these types of behaviour into a functional combination (De Bruin, 1980). Lesions to the nerves that relay chemosensory information from the pharyngeal region to the brain also caused changes in the pattern of fanning and sexual behaviour, which suggests that chemosensory cues from the eggs play a significant role in determining the relative balance between parental and sexual behaviour (Segaar *et al.*, 1983). Experiments have suggested that a parental male will move into a compartment of an aquarium into which water is run that had previously run over fertilised eggs (Golubev and Marusov, 1979).

Males castrated early during their parental cycle showed an almost normal fanning cycle and an increase in aggression when the eggs hatched (Baggerman, 1968), but they did not start another reproductive cycle as an intact male could do. The pituitary hormone prolactin may play a role in the control of fanning behaviour. The increase in the time spent fanning during the parental cycle was correlated with an increase in prolactin cell activity assessed by quantitative electron microscopy techniques (Slijkhuis, 1980), and low doses of mammalian prolactin can increase the amount of fanning in the sexual phase (Molenda and Fiedler, 1971). Prolactin is of course closely associated with parental behaviour in mammals, but in teleosts its major role seems to be in osmoregulation (Chapter 2).

Male Sneaking and Nest Raiding

A male's nest is not only the focus of his activities, it is also a focus of interest to neighbouring males. In *G. aculeatus*, males will attempt, sometimes with great perseverance, to approach the nest of another male. When approaching another's nest, the male adopts behaviour that can best be described as 'sneaking'. On reaching the territory border, he sinks to the bottom and his breeding colours tend to fade. He approaches close to the nest to be raided by a series of slow glides, sinking on to or close to the bottom between glides. The last short distance to the nest is covered in a rapid dash. If the resident male detects and attacks the sneaking male before the nest is reached, the sneaker flees, but if the sneaker succeeds in reaching the nest, the owner has considerable difficulty in driving the sneaker off. Although sneaking was first described in a laboratory study (van den Assem, 1967), it occurs frequently in natural populations (Wootton, 1971b; Kynard, 1978a).

What a sneaking male does, once he has reached the nest, depends on the phase the male whose nest is raided has reached. If the raided male has no eggs in his nest, the sneaker will tear at the nest, destroying it. He may even attempt to carry nest material back to his own nest (Wootton, 1971b, 1972; Li and Owings, 1978a). Sneaking occurs at a high frequency during courtship, with the sneaker either preventing the female from entering the nest by lying across it or in its entrance, or following the female through the nest before the resident male so that the sneaker steals a fertilisation. The sight of a courtship seems to be a potent situation for eliciting sneaking (Li and Owings, 1978a,b). When the raided male has eggs in the nest, the sneaking male attempts to steal the eggs and either eats them or carries the stolen eggs back to his nest. In most cases, the stolen eggs are eaten, although in some cases the raiding male may steal eggs that he has previously fertilised by sneaking (van den Assem, 1967; Black, 1971; Wootton, 1971b; Kynard, 1978a). The early studies suggested that sneaking and egg stealing were only perpetrated by males without eggs in their nest but this is not generally true. The males of Lake Wapato were enthusiastic sneakers and egg stealers and showed this behaviour at all stages of their reproductive cycle. Eggs that were stolen were invariably eaten and sneaking to steal fertilisations was not seen. Males that were guarding eggs that had reached the eyed stage had the highest number of eggs in their stomachs. The eggs in the stomachs were usually not at the same stage as their own eggs, so males seemed to be capable of stealing and eating other eggs while showing parental behaviour towards their own. Although fry were rapidly eaten by other males if the parental male was removed, only a small proportion of males had fry in their stomachs, which suggests that male defence of fry was especially effective. Predation was greatest on eggs in the pre-eyed stage. In addition to egg predation by sneaking males, groups of females and non-reproductive males would also raid nests and eat the eggs. The resident male had considerable difficulty in driving off such packs of raiders (Kynard, 1978a). When a sneaking male succeeds in raiding a nest, the activity of the sneaker and the resident male frequently attracts other males to the nest so that a group raid occurs (Black and Wootton, 1970; Wootton, 1971b).

Since a male might both be a sneaker and the victim of sneaking, there must be some net benefit gained from this behaviour, which at first sight seems to disrupt reproductive behaviour so much. In other groups of fish, sneaking and stealing fertilisations have been described, but in those cases, it is usually small males that have not built a nest or do not hold a territory that sneak and steal (Gross, 1984). It represents an alternative strategy for achieving reproductive success. In sticklebacks, the sneakers are not smaller, they are usually territorial and have nests. However, male sticklebacks are in competition. They compete for a suitable territory, they may have to compete for material with which to build a nest, and they compete for gravid females. Furthermore, because their territorial behaviour restricts them to a limited area, they may also be relatively short of food if the area in which the territories are situated is not productive. Sneaking and nest raiding provide a behavioural mechanism by which a male can compete for the resources which are in short supply. The nest of a neighbour provides a supply of nest-building material, it provides a site at which eggs can be fertilised, and the eggs in another male's nest can be eaten to subsidise the energy costs of a raiding male's own reproductive attempt. A raid may also cause a neighbour to revert to the nest-building phase and thus allow the raider more opportunity to court the available gravid females. Li and Owings (1978a) found that raiding tended to occur when the males involved were relatively equal in dominance status. A dominant male would disrupt the courtship of a subordinate male by a direct attack. Although sneaking and raiding seem to be potent forms of inter-male competition, the circumstances which determine when, and the form the raiding takes, have yet to be satisfactorily analysed.

Nest raiding and stealing fertilisations also occur in *P. pungitius*, *G. wheatlandi*, *C. inconstans* and *A. quadracus*, so it is a form of intra-male competition which is common to the family, though whether it reaches the intensity that has been recorded for *G. aculeatus* is unclear.

Behavioural Correlates of Reproductive Success

The reproductive behaviour of male *G. aculeatus* has been described in sufficient detail both in experimental and field conditions for some of the behavioural factors that are correlated with reproductive success to be identified. Two important factors are territory size and the concealment of the nest.

Territory Size and Reproductive Success in Gasterosteus aculeatus. Territory size can influence success in both courtship and rearing the eggs. Solitary males forced to accept territories with dimensions less than 300 X 300mm were less successful at getting females to enter their nests than solitary males with larger territories. Even when the males with the small territories were eventually successful, the courtship took three times longer than for males with larger territories. The courtship was also more likely to break down once the female had swum down to the male's nest. Males with small territories tended to be more aggressive towards the female; they also tended to spend more time fanning at the nest during the courtship. Males with large territories also had an advantage in courtship when several territorial males were present. Overall, the chances of a successful courtship were less and the duration of the courtship longer in the situation where rival males were competing for the attention of the female. However, the males with the largest territories were more successful at getting females down to their nest than would be expected if all the males had equal chances. Once at the nest, the female was more likely to enter the nest of a male with the largest territory (van den Assem, 1967).

When males were kept in groups, the males that had the largest territories c ' ⸱ ᴌ the parental phase were more successful at rearing eggs than males with smaller territories. During the course of the parental phase, the territory size of the successful males declined to that typical of the unsuccessful males (Black, 1971).

In both these experimental studies, the males that initiated the highest number of aggressive encounters tended to have the largest territories. Inter-nest distance between males in Lake Wapato was positively correlated with the level of aggression of the males (Kynard, 1978a).

Curiously, in the multi-species associations studied in salt-marsh pools in the St. Lawrence Estuary, FitzGerald (1983) found that there was a negative correlation between inter-nest distance and the number of eggs in *G. aculeatus* nests. There was also a negative correlation between frequency of aggression and the number of eggs in the nest in *G. aculeatus*, *G. wheatlandi* and *P. pungitius*. These results show that further analysis of the relationship between aggression, territory size and reproductive success is required. Over what range of territory sizes are males successful at producing young and what is the best territory size for a male in relation to the density of con- and hetero-specific competitors, the amount of cover and other environmental factors?

Nest Concealment and Reproductive Success. Observations on natural populations of *G. aculeatus* indicate that nests built in or close to shelter are more likely to contain fry than nests in the open (Moodie, 1972a; Kynard, 1978a). Experiments suggest that this is at least partly due to a reduction in interference by rival males. In an experiment in which groups of males were kept in small paddling pools, the males that nested in flowerpots were reproductively more successful than males that nested in the open. Males with a nest in a pot spawned earlier and more often. Their success in hatching eggs was greater and the variation in hatching success was less. In comparison with males with nests in the open, the males in pots suffered fewer stolen fertilisations and fewer nest raids and were involved in fewer territorial encounters. Even if the male nesting in a pot was separated from a rival by a glass screen, he was still more successful. Males in pots had less variable lengths to their fanning bouts than males with nests in the open, which may help to account for the pot effect when males could not be interfered with by a rival (Sargent and Gebler, 1980). Males with and without pots did not differ in territory size, in the total number of courtships in which they engaged, nor in the number of times a female rejected the nest having followed the male to it. When the performance of the same males with and without pots was compared, a given male attracted more nest choices by females and suffered fewer intrusions from rivals when his nest was in a pot. This suggests that female choice depends on the quality of the territory more than on the quality of the male (Sargent, 1982). One interpretation is that females chose males that were least likely to have their courtship interfered with by rival males. Sargent suggested that with the dorsal-pricking sequence a male 'parades' a gravid female close to the territorial boundaries. Creeping through then stimulates any rivals to raid the nest, but the raid is premature because the female is not in the nest and so the raider fails to steal a fertilisation. Clearly, this functional interpreation is not incompatible with the motivational interpretations described earlier.

Nest concealment will not only be effective against raiding conspecifics, it may also reduce heterospecific predations on eggs and fry (Moodie, 1972a).

Other Correlates of Reproductive Success. Once a female has followed a male down to his nest, she pokes her snout into the nest entrance. At this point, the courtship frequently breaks down and the female withdraws, often chased off by the male (Wootton, 1974b). The chances of the female withdrawing were found to be less if the nest already contained some eggs, that is, the male had either been successful in previous courtships or had successfully stolen eggs during a raid (Ridley and Rechten, 1981). This behaviour may have the advantage that if some of the eggs in the nest are predated, the risk is shared between the eggs of more than one female (Rohwer, 1978; Ridley and Rechten, 1981). Males with eggs in the nest are also less likely to abandon the nest in the presence of a predator (Pressley, 1981). Rohwer (1978) speculated that fanning during courtship may be an attempt by a male to signal to the courting female that he has some eggs already in his nest, even though this may not be true.

Quantitative Aspects of Stickleback Reproduction

Size of Propagules

The diameters of the eggs of sticklebacks are typically between 1.1 and 1.7 mm, though *S. spinachia* has larger eggs with a diameter of about 2 mm. At a site in the St. Lawrence Estuary where the four species were sympatric, *P. pungitius* had the smallest eggs (diameter 1.14 mm) and *A. quadracus* the largest (diameter 1.40 mm), with *G. wheatlandi* (1.2 mm) and *G. aculeatus* (1.39 mm) falling between (Craig and Fitz-Gerald, 1982). At Woods Hole, further to the south, the ovulated eggs of *A. quadracus* were significantly larger than those of *G. aculeatus*, the mean diameters being 1.62 mm and 1.31 mm, respectively (Wallace and Selman, 1979). Stickleback eggs are similar in size to those of other summer-spawning fishes of fresh and brackish waters of the northern hemisphere (Wootton, 1984), so sticklebacks have not compensated for their relatively small body size at sexual maturity by producing relatively small eggs. At hatching, the fry are about 5 mm in total length. In *G. aculeatus*, a positive correlation was found between egg diameter at fertilisation and the length of the fry at hatching (D.A. Fletcher and R.J. Wootton, unpublished).

Both the period of spawning of sticklebacks and the size of their propagules are similar to those of the majority of fishes in fresh or brackish water in the same geographical regions. This highlights an urgent need for studies of the comparative ecology of fishes in their first few months of life to determine how the species interact and the mechanisms that permit their coexistence at this early period in their life history.

Fecundity of Sticklebacks

Because sticklebacks can spawn several times in a breeding season, different measures of fecundity are available. The most important are fecundity per spawning, fecundity per breeding season and fecundity per lifetime. Most published estimates are only for fecundity per spawning. Within a species, fecundity per spawning is a function of body size. The relationship with body length is curvilinear and can be described by a function of the form:

$$F = aL^b$$

or in the linear form:

$$\ln F = \ln a + b \ln L$$

where F is fecundity, L is body length and a and b are parameters that can be estimated by regression analysis (Wootton, 1979). For females belonging to the completely, partially and low plated morphs of *G. aculeatus* living in the Little Campbell River, the exponent, b, took the value of 2.43. The value of a did vary significantly between the morphs so that the expected fecundity of a female of length 50 mm was 133 for the low plated, 154 for the partially plated and 163 for the completely plated morph (Hagen, 1967; Wootton, 1973b). For this species the fecundity typically ranges from about 40 to 400 eggs, depending on the size of the female (Figure 7.7).

A comparison of the fecundities of four sympatric species at a site in the St. Lawrence River showed that *G. aculeatus* had the highest mean fecundity of 366, *A. quadracus* the lowest, 36, with *G. wheatlandi* and *P. pungitius* similar with fecundities of 80 and 76, respectively. These differences partly reflected differences in the average body size of the four species, but a unit increase in body weight produced a far greater increase in fecundity in *G. aculeatus* than in *A. quadracus*, with the other two species intermediate (Craig and Fitz-Gerald, 1982). When fecundity was recalculated as the total volume of eggs per spawning and expressed per unit of somatic weight, the two *Gasterosteus* species were similar, but *P. pungitius* and *A. quadracus* were much lower.

There are no estimates of breeding season fecundity for females in a natural population in any stickleback species. In an experimental group of *G. aculeatus* females kept over a single breeding season, the average number of eggs produced by fish fed a high ration was 973 (D.A. Fletcher and R.J. Wootton, unpublished). In many populations of *G. aculeatus* and probably in the other species too, the majority

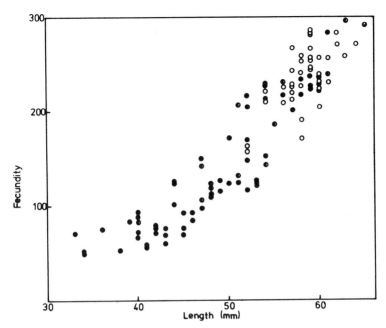

Figure 7.7: Relationship between Fecundity per Spawning and Length for Female *G. aculeatus* from Little Campbell River, British Columbia. Closed Circles, Low Plated Morph; Half Circles, Partially Plated Morph; Open Circles, Completely Plated Morph. Data from Hagen (1967)

of females die during or after their first breeding season (Chapter 9), so that their lifetime fecundity is the same as their breeding season fecundity. Estimates of both these fecundities are urgently required for all the species. Without this information, it is impossible to interpret the significance of the differences in the fecundity per spawning which the species exhibit. It is possible that the species with a low fecundity per spawning, such as *A. quadracus*, compensate by having a shorter interval between successive spawnings and a longer breeding season (Wallace and Selman, 1979; Craig and FitzGerald, 1982).

Fecundity of a male can be defined as the number of eggs he fertilises per parental cycle, per breeding season or per lifetime. It is not clear if there is a maximum number of eggs a male can successfully rear to hatching. In laboratory studies of *G. aculeatus*, up to seven clutches have been successfully reared by a male. Some males in Lake Wapato had over 2000 eggs in their nest. In this lake, the average number of eggs found in a nest declined significantly by about 200 eggs per month between May and August. Males towards the end of the breeding season were far less successful at rearing the eggs through to fry and they had a tendency to desert their nest during the parental cycle (Kynard, 1978a,b). The causes of these changes in egg number and male parental effectiveness are not known.

During a breeding season, a male that re-nested immediately after completing a parental cycle could complete up to about seven cycles, but the males in Lake Wapato rarely re-nested immediately and only a small proportion were observed re-nesting at all (Kynard, 1978a). The breeding season fecundity of males has not yet been measured. Kynard (1978a) drew attention to the tremendous reproductive potential of the stickleback population in Lake Wapato. He collected 485 nests from limited sampling areas in the lake and counted 290 500 progeny in these nests. Especially early in the breeding season, the males were rearing a high percentage (76.8 per cent) of their eggs through to fry, so that on average each parental male was producing over 400 free-swimming fry. The population dynamics of sticklebacks are considered further in Chapter 9.

Energy and Time Budgeting of Stickleback Reproduction

Because the reproductive investment of the females is primarily in the form of eggs, which can be collected and weighed and their energy content measured, it has proved relatively easy to measure the energy costs of reproduction for the female. The complex behavioural investment of the male has proved less easy to quantify in terms of energy expenditure, but preliminary budgets of the expenditure of time are possible.

Energy Budget of Female. Only female *G. aculeatus* have yet been studied, but the differences in fecundity per spawning and the length of the breeding season show that comparable studies on the other species are urgently required.

Until the start of the breeding season, the ovaries form only a small proportion of the total body weight and between autumn and spring they increase in size and energy content only slowly. For populations of *G. aculeatus* living in a small lake and a river in Wales, it was estimated that up to March less than 2 per cent of the energy content of the daily food consumption was accumulated in the ovaries. Over this period, the energy costs of maintenance and activity accounted for virtually the whole of the energy expenditure (Chapter 5). However, compared with somatic growth, the growth of the ovaries seemed to have a relatively high priority because even in periods when there was a loss in the energy content of the soma, the ovaries maintained or even increased their energy content (Wootton *et al.*, 1980a).

During the breeding season, the eggs from each spawning and the total weight of food consumed between successive spawnings can be measured in the laboratory. The energy content of eggs and food can be determined by bomb calorimetry. Using these techniques on females fed ad lib with *Tubifex*, Wootton and Evans (1976) estimated that on average for every 100 J of food consumed in the interval between the first and second spawnings, 26 J of eggs were produced at the second spawning, and other outputs were 3 J of somatic growth and 11 J of faeces. The energy content of the eggs was estimated to be 22.6 kJ g^{-1} dry weight, so that on average each egg contained 7-8 J at fertilisation. During the interval between successive spawnings a female ate, on average, about her own body weight of *Tubifex*, although there was a significant drop in food intake in the 24 h before spawning. The usual time between spawnings was 3 to 5 days.

In those experiments, the female was in control of her food consumption because food was always present, but what happens to the investment in egg production when the food supply to the female is restricted? The number, weight and energy content of eggs produced at a spawning are primarily determined by the size of the female, so once a female becomes committed to spawning, the energy expenditure in the form of eggs has to be met whatever the food consumption. The effect of this is that at low rations, the energy content of the eggs accounts for a high percentage of the energy of the food consumed, but the energy costs of egg production are subsidised from the female's body so she loses weight. A female on low rations can lose about 20 per cent of her body weight subsidising egg production (Wootton, 1977; D.A. Fletcher and R.J. Wootton, unpublished). For a non-breeding female a daily ration equivalent to about 2 per cent of her body weight is sufficient for her to maintain a constant body weight (Allen and Wootton, 1982a), but during the breeding season a female has to consume a daily ration of about 8 per cent of her body weight if she is not to lose weight between successive spawnings (Wootton, 1977). It is possible that during the breeding season the feeding motivation of the female does increase, so that she attempts to match her energy income to the increased energy expenditure associated with egg production.

For the female, the time spent in courtship is relatively brief, a few minutes at intervals of about a week, so neither the time nor the energy cost of the female's reproductive behaviour is likely to be significant. A female is likely to maximise her reproductive success if she can maximise the time she spends foraging to meet the energy costs of egg production and maximise the number of spawnings she can achieve in a breeding season.

Time Budget of Males. Reproductively active males face a different problem. Their reproductive behaviour, particularly during the parental phase, will occupy a large proportion of the available time, yet they are restricted to a small territorial area for foraging. They do not fast during the breeding season (Feuth-de Bruijn and Sevenster, 1983; Stanley, 1983), but it is not known what priority is given to foraging by a reproductively active male. Observations on both a natural population and groups of males in the laboratory suggest that once the territorial system has been established, the male may spend relatively little time in active defence (Wootton, 1972; Stanley, 1983), and he may spend long periods relatively inactive until a parental cycle is established.

During the parental phase, there is evidence from *G. aculeatus* that the time budgeting of the male does have an effect on his success in hatching young. Development of the eggs is favoured if the male can fan in relatively long bouts (van Iersel, 1953; van den Assem, 1967), or if the male can reduce the variation in the length of the fanning bouts and the inter-bout interval (Sargent and Gebler, 1980). A male is more likely to achieve this if his choice of nest site or his territorial defence has reduced the chance of raids on his nest.

Effect of Food and Population Density on Reproductive Success

Given that reproduction does impose energy costs over and above those that must be met by a stickleback in non-reproductive condition, the effect of food availability on reproduction is going to be important. But even if the food supply is adequate, the density of fish will also be important because to be successful, a male must obtain a territory of at least a minimum size. This implies that in a given area there is a maximum number of territorial males that can be accommodated.

Effect of Food and Density on Females. The food availability before the start of the breeding season has two effects on reproductive success. Female fecundity is a function of body size. The growth rate is strongly correlated with food availability (Chapter 6), so the body size of the female as she enters a breeding season and hence her fecundity per spawning will be related to her prior history of food consumption. Secondly, the proportion of females in a population that reach sexual maturity is related to their food supply (Wootton, 1973a). Sticklebacks have to reach a critical size before they become sexually mature. This critical size varies from population to population, and the variation has a genetic component (McPhail, 1977), but females that, because of inadequate growth, fail to reach that size at an appropriate age will not reproduce. In some populations a proportion of females reach sexual maturity at 1 year old whereas others do not reach it until they are 2 years old (Chapter 9).

Once they have started breeding, the number of spawnings by a female and hence the breeding season fecundity depend on the food supply. The results of an experiment in which female *G. aculeatus* were fed enchytraeid worms at a fixed ration during the breeding season are shown in Table 7.1.

Table 7.1: Effect of Food Ration on Breeding Season Fecundity of Female *G. aculeatus*

Ration as % of body weight at start of breeding season	Mean fecundity $\pm$ SE
2	208 $\pm$ 48.0
4	331 $\pm$ 108.8
8	550 $\pm$ 171.9
16	973 $\pm$ 152.7

If gravid female *G. aculeatus* are kept in a confined space, they will be aggressive, and the most aggressive females can prevent others becoming mature or cause them to abort their clutches (Wootton, 1973a; Li and Owings, 1978a). It is not known whether these effects occur in natural populations. Females are usually observed foraging in schools and there are no reports of inter-female aggression in such schools (Black and Wootton, 1970; Kynard, 1978a; FitzGerald, 1983).

Effect of Food and Density on Males. There is, as yet, no evidence that larger males are at an advantage when competing for territories or that females prefer to mate with larger males, although in other teleosts such effects have been found (Downhower and Brown, 1981). Low food rations do tend to inhibit nest building, an effect described both in *G. aculeatus* and in *C. inconstans* (Stanley, 1983; Smith, 1970).

Sexually mature *G. aculeatus* kept on a ration of 2 per cent of body weight had significantly smaller kidneys than males on rations of 6 and 18 per cent of body weight. When male *G. aculeatus* were kept in groups of 3, 6 or 9 in 1800 X 370 X 250mm tanks and fed either 2, 6 or 18 per cent of their body weight, both ration and density were found to have significant effects on territoriality and nest building. At the lowest ration fewer males and a lower proportion of males in a group could hold a territory and build a nest over the same period of time. Even at the lowest density and highest food ration, not all the males were able to obtain territories. When the males were put in aquaria on their own, a significantly higher proportion of males at the 6 and 18 per cent rations then built nests, whereas only a small proportion of the males on the 2 per cent ration built nests even when isolated from other males (Stanley, 1983). At high rations some males were inhibited from setting up territories and building nests by the behaviour of their competitors, although when tested in isolation they showed that they were physiologically capable of building nests. At the low rations, even in isolation some males were physiologically incapable of building a nest. Although at higher densities more males took up territories and built nests, the proportion of males that were successful was inversely related to male density. A similar inverse relationship between male density and the proportion acquiring territories was found by van den Assem (1967). Thus, for the males in a given area, the proportion that succeed in gaining a territory and building a nest is directly related to food supply but inversely related to male density. More males can be accommodated in a given area if they take up their territories simultaneously rather than sequentially (van den Assem, 1967).

Ration also had a significant effect on the fanning behaviour of isolated male *G. aculeatus*. Each male received a ration of either 2, 6 or 18 per cent of body weight and was allowed to fertilise one clutch of eggs. The males on the lowest ration showed a significant reduction in the time they spent fanning, but they were not significantly less successful at hatching their clutch (Stanley, 1983). Rohwer (1978) speculated that a male stickleback might cannibalise some of the eggs in his nest to meet the energy costs of the current parental cycle and to keep the male in a condition to start another cycle. The evidence from this experiment and Kynard's (1978a) field observations do not support his speculation.

The effects of food and density on both the male and female that have been demonstrated experimentally suggest that at high food levels and low densities, reproductive success will be high, whereas at high populations densities or at low food levels, success will be low. These responses should enable the sticklebacks to track changes in food availability closely and rapidly. Because the young fish quickly start exploiting the same types of food as the parental generation (Chapter 4), the food experienced by the parental generation during their breeding season may be a good predictor of the food level that will be experienced by their offspring.

Conclusion

The reproductive biology of the sticklebacks is dominated by the differences in the roles of the female and the male. A female provides each offspring with large (at least relative to her own size) provisions to meet the costs of the first few days of life. But she plays no role in protecting her offspring. The male provides his offspring with a relatively safe, concealed place in which to spend the first and most vulnerable days of their lives, a place which he defends from intruders at a risk to himself. Maynard Smith (1977) used game theory to analyse the conditions under which male parental care might evolve. It would be favoured if the female had to use up resources in parental care that could be devoted to egg production without a compensatory gain in the survival of the eggs when both male and female showed parental care. Several aspects of the biology of sticklebacks favour the male assuming the parental role. The female, freed from parental duties, can forage widely to obtain the food required to meet the costs of the production of further clutches of eggs. She also avoids paying the energy costs of ventilating the eggs. These costs have not yet been measured, but are likely to be significant. If her movements were restricted and she was paying the cost of parental behaviour, her breeding season and lifetime fecundity would be considerably reduced. The male, by his territorial behaviour, sequesters a suitable spawning site which will probably improve his chances of mating with several females. His courtship behaviour and the use of the nest, together with external fertilisation of the eggs, ensure that there is a high probability that he is the father of the eggs in his nest (Maynard Smith, 1982).

Because a male has to attract a female down to his nest, at a time when other males are attempting to attract the same female, there will be inter-male competition. This competition is reflected in the well developed male breeding colours in some species, in the fighting to acquire a territory with a suitable nesting site, and in the subsequent mutual interference between males in the form of sneaking and nest raiding. Once a high proportion of males in a population have acquired eggs and are in the parental phase of the sexual cycle, a gravid female can find herself in competition with other females for the attentions of a courting male. This inter-female competition may be reflected in the prominent, distended and usually silvery belly the female develops just prior to spawning. There is less evidence, at least from observations on natural populations, that there is significant inter-female fighting to obtain access to males.

Some of the differences in the reproductive behaviour of the stickle-back species can be convincingly related to differences in their defence against fish predators. *Gasterosteus* is morphologically better protected against predators than *Pungitius* or *Culaea* (Chapter 8). Correlated with this protection, *Gasterosteus* will nest away from dense vegetation and adopt conspicuous breeding colours, especially prominent in *G. aculeatus*. This species is also the most aggressive of the sticklebacks and this high level of aggression probably explains the dorsal-pricking and nest-return components of the male's courtship. He has to defend a nest that is often relatively exposed to both conspecific and heterospecific predators, but *G. aculeatus* also seems to be most bold towards poten-tial predators (Huntingford, 1976a).

8 INTER-SPECIFIC INTERACTIONS

Predation

A disadvantage of the small adult size of the sticklebacks is that through-out their lifespan they are the potential prey of a wide variety of predators. Although there is considerable qualitative information on the ecology of predation on sticklebacks, there is inadequate quanti-tative information so that the impact of predation on stickleback populations has still to be assessed. This lack of quantitative informa-tion is particularly unfortunate because evidence is accumulating that predation is probably a major selective factor in the evolution of the extensive intra- and inter-population variation found in several stickle-back species (Chapter 10).

Predators of Sticklebacks

Invertebrate Predators

A few carnivorous invertebrates are sufficiently large to be predators of sticklebacks in their first few months of life. Leeches eat stickleback eggs: the North-American species *Haemopsis marmorata* can consume about 100 eggs of *G. aculeatus* at one feeding (Moodie, 1972a). A male stickleback guarding a nest containing eggs will remove leeches, snails or other invertebrates that get too close to the nest. Once the eggs have hatched, the small fish become the potential prey of large insects such as the nymphs of dragonflies (Odonata), carnivorous water bugs such as *Notonecta*, *Ranatra* and *Lethocerus* (Hemiptera) and the larvae or adults of *Dytiscus* (Diptera). These predators either ambush or stalk their prey and so tend to operate in close physical proximity to their victims. The prey, once grasped, are not usually manipulated or mouthed extensively, which reduces the effectiveness of defensive apparatus such as spines (Reist, 1980b). Invertebrate predation tends to be strongly size selective. In experiments with *G. aculeatus* as prey, nymphs of the dragonfly *Aeschna* predated 98 per cent of sticklebacks in the size range 15-25 mm, but took only 4.4 per cent of fish 30-40 mm long and 1.1 per cent of those 50-60 mm in length (Reimchen, 1980). Similarly, the ambush predator *Ranatra* tended to prefer the relatively smaller fish from a group (Hay, 1974). Both these experimental studies were complemented by field studies which suggested that the invertebrates were significant predators on sticklebacks.

Vertebrate Predators

Mammals. Both mink (*Mustela vison*) and otters (*Lutra lutra* and *L. canadensis*) include sticklebacks in their fish prey, although usually as only a small component of their diet (Wootton, 1976). In north-eastern Alberta, *C. inconstans* was important in the diet of mink and otters from a habitat dominated by lakes (Gilbert and Nancekivell, 1982).

Birds. A variety of fish-eating birds including gulls, terns, ducks, grebes, kingfishers and herons eat sticklebacks (Wootton, 1976) and for some of these species, at least in some habitats, sticklebacks are the major food item.

Arctic terns (*Sterna paradisaea*) nesting in an archipelago off the coast of Finland fed heavily on *G. aculeatus* during the breeding season and also fed them to their young. Common terns (*S. hirundo*) nesting in the same area also had *G. aculeatus* as their commonest prey but fed more extensively on other fish species than did the Arctic tern. The latter species tended to select prey that were between 50 and 55 mm in length, although in May the mean size of the sticklebacks was significantly smaller than that. The food of the newly hatched terns was smaller than that of the adults, but once the fledglings were about a week old they were fed sticklebacks that were as large or even larger than those eaten by adult Arctic terns. It was estimated that during the 23 days or so that it took a young tern to fledge, each chick consumed about 750 g of food, by far the greatest proportion of which consisted of sticklebacks (Lemmetyinen, 1973).

In Boulton Lake on the Queen Charlotte Islands, British Columbia, sticklebacks are the prey of several species of birds including the common loon, *Gavia immer* and the belted kingfisher, *Megaceryle alcyon*. A pair of loons that nested on the lake was estimated to take about 75 000 sticklebacks each during the 6 months they spent at the lake, and each kingfisher was estimated to take about 5000 fish during a 7-month stay. The fishing habits of the two species suggested that loons were feeding predominantly on adult females whereas the king-fishers, which foraged close to the shoreline, were feeding on adult males. During the winter, grebes and mergansers foraged on the lake (Reimchen, 1980). The merganser, *Mergus serrator*, fed almost exclusively on *G. aculeatus* in Lake Myvatn, Iceland (Bengtson, 1971).

The night heron, *Nycticorax nycticorax*, fishes in the tidal rivers and salt marshes of the St. Lawrence Estuary where four species of stickleback breed. Although *G. aculeatus*, *G. wheatlandi*, *P. pungitius* and *A. quadracus* are present, the herons seemed to be feeding only on *G. aculeatus*. The reasons for this apparent specialisation on only one of the species are not known, although this differential predation may be one of the factors that permit the coexistence of the four species in the area (FitzGerald and Dutil, 1981).

Reptiles. In parts of North America, the water-loving garter snakes *Thamnophis* are found in the same habitats as sticklebacks. In southern California, *T. couchi hammondi* is a predator on *G. aculeatus* (Bell and Haglund, 1978), and in other areas garter snake predation on stickle-backs may be important.

Fish. Freshwater piscivorous fish that eat adult sticklebacks include trout (*Salmo* spp.), char (*Salvelinus* spp.), pike (*Esox*), perch (*Perca*) and squawfish (*Ptychocheilus*). Small sticklebacks may be eaten by sculpins (*Cottus* spp.). In the sea, sticklebacks have been found in the stomachs of cod (*Gadus morhua*) and other gadoids. These predators display a variety of hunting techniques. The salmonids and perch pursue the sticklebacks. Once the prey is caught, it is manipulated in the mouth, but during this manipulation the stickleback has a chance to escape. Squawfish pursue only over short distances and capture of the stickleback is aided by strong oral suction. The stickle-back is easily swallowed so that once engulfed it has little chance of escape. The pike is an ambush predator, lying virtually motionless in vegetation until the prey comes close, then lunging rapidly forward, seizing the prey in the large mouth with its formidable array of teeth (Figure 8.1) (Hoogland *et al.*, 1957; Moodie *et al.*, 1973).

Carnivorous fish have to reach a certain size before they can cope with fish prey. Experiments in which pike 150-250mm in length were allowed to prey upon *G. aculeatus* and *P. pungitius* suggested that the spines of the sticklebacks posed problems for the predator (Figure 8.1) (Hoogland *et al.*, 1957). In natural populations of pike, *G. aculeatus* tend to form a significant portion of the diet only when the pike

Figure 8.1: Small Pike (*Esox lucius*) Manipulating *G. aculeatus*. After Hoogland *et al*. (1957)

has exceeded 200mm in length. In Lake Windermere (UK), only for pike longer than 400mm did sticklebacks form a larger proportion of the diet than species that lacked spines. Brown trout (*S. trutta*) include sticklebacks in their diet once the trout have reached a size of about 300mm (Wootton, 1976). There was a positive correlation between the size of *G. aculeatus* eaten and the size of cut-throat trout (*S. clarki*) eating them in Meyer Lake (Queen Charlotte Islands). Trout of 200mm standard length typically took sticklebacks with a length of 30-40mm and those of 400mm took sticklebacks 90mm in length (Moodie, 1972b). In Lake Wapato (Washington State), *G. aculeatus* were eaten by rainbow trout (*S. gairdneri*). A far higher percentage of trout had sticklebacks in their stomachs in winter than in spring, because in spring only trout larger than 350mm fed on sticklebacks whereas in winter trout as small as 220mm were eating them (Hagen and Gilbertson, 1973a). There have not yet been systematic studies of size selection by the fish predators of sticklebacks, so it is not clear to what extent different size classes experience relatively low or high predatory pressures.

Defences against Predation in Sticklebacks

Given the variety of their natural enemies, it is not surprising that
sticklebacks have a range of defences against their predators. The
main defences are morphological adaptations, especially the spines,
the adoption of cryptic coloration and behavioural adaptations.
Defence against predators can be classified as primary or secondary
(Edmunds, 1974). Primary defence reduces the possibility of detection by the predator and secondary defence reduces the chance of
being eaten once detected by the predator.

Morphological Adaptations

Morphological adaptations act as secondary defences by increasing
the chance of the stickleback escaping if captured by a predator
and by reducing the chance that the predator will attack a stickleback
once detected. In comparison with many other small fishes, sticklebacks have much more substantial and robust skeletons and this makes
them 'durable' so that they may often be relatively unscathed if they
can escape after capture (Benzie, 1965; Moodie, 1972b). Both the
extent of the body armour in the form of bony plates and the size of
the spines varies between species and within species. The completely
plated morph of *G. aculeatus* represents probably the most heavily
armoured stickleback, and *C. inconstans*, which lacks developed plates
and has small spines, is the least armoured stickleback. But in freshwater populations of *G. aculeatus*, *P. pungitius* and *C. inconstans*,
there are populations in which the spines are reduced or lost and the
lateral plates few or absent (Chapter 10).

When viewed from the front, the stickleback with its dorsal and
pelvic spines erect looks an uncomfortable meal (Figure 8.2). In *G.
aculeatus*, the erection of the spines can increase the cross-sectional
area of the trunk region by about 75 per cent (Gross, 1978) and the
spines are sharp, as anyone who handles sticklebacks will quickly
discover. In some populations, the spines may also be serrated. The
function of these serrations is not clear, but they may act as a cutting
edge. In *P. pungitius* and *A. quadracus*, the dorsal spines alternatively
incline to the right and left of the midline, which increases the area
of the dorsal region protected by spines. Although the spines of *P.
pungitius* are relatively smaller than those of *Gasterosteus*, the number
and inclination mean that the back of the fish presents a high density
of spines to any potential predator.

An important feature of the spines is that they can be locked in the

Figure 8.2: Front Views of Sticklebacks with Spines Erect. Left, *G. aculeatus* (from River Rheidol); Middle, *P. pungitius* (from St. Lawrence Estuary); Right, *Apeltes quadracus* (from St. Lawrence Estuary)

erect position, so that if the stickleback becomes fatigued the spines are not depressed. The mechanism that locks the spines ensures that pressure exerted on them does not cause their collapse; a predator must be able to ignore or break them (Hoogland, 1951).

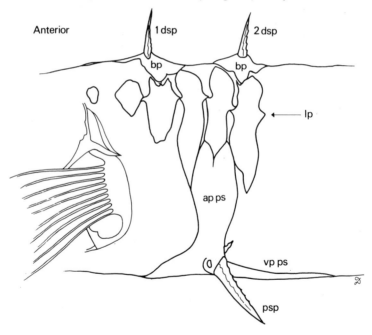

Figure 8.3: Thoracic Region of a Low Plated *G. aculeatus* Collected from River Rheidol, Wales. Drawn from Alizirin-red Stained Specimen. bp, Basal Plates of Dorsal Spines; dsp, Dorsal Spines; lp, Lateral Plates; apps, Ascending Process of Pelvic Skeleton; vpps, Ventral Process of Pelvic Skeleton; psp, Pelvic Spine

The dorsal and pelvic spines, the large, bony pelvic and pectoral skeletons, and the anterior lateral plates form a tank-like armoured thoracic region which is particularly well developed in *Gasterosteus* (Figure 8.3) (Gross, 1978; Reimchen, 1983), but most weakly developed in the largest of the sticklebacks, *Spinachia*. This latter species also has the smallest spines, although those in the dorsal row are numerous.

Both experimental and field studies have suggested that the spines are effective as a defence against fish predators. Experiments in which perch or pike were used as predators showed that when *G. aculeatus* and *P. pungitius* were presented together with a similarly sized but spineless species such as the minnow (*Phoxinus phoxinus*), the predators took the minnows first, then the *P. pungitius* and only finally the *G. aculeatus*. Even when the prey were presented singly, this order of preference was still displayed by the predators. If the sticklebacks had their spines cut off, they were treated as spineless fish by the predators. The similarity in preferences of the perch and pike holds despite the differences in their hunting technique, though captured sticklebacks were more likely to escape from the perch than from the pike (Hoogland *et al.*, 1957). The latter is a specialised piscivore, whereas the perch is a generalist carnivore. In these experiments, the pike were relatively small, 150-250 mm long, and descriptions of the natural diet of pike suggest that sticklebacks do not form an important part of their diet until the pike have reached over 200 mm in length.

An experiment in which pike were allowed to prey on *C. inconstans* gave less clear-cut evidence of the effectiveness of spines. Although a higher proportion of fish that had had their pelvic spines clipped were eaten by the pike than unclipped fish, the difference was only significant for the prey in the size range 30.0-39.9 mm (Reist, 1980a). Since both groups still had their dorsal spines, the importance of this result is not clear. Sticklebacks with their spines clipped could also show abnormalities of behaviour which make them more susceptible to predation, an effect that needs to be considered in the design of experiments on predation.

Broad geographical surveys of freshwater populations of *G. aculeatus* along the west coast of North America and in Europe show that where predatory fish are common, the sticklebacks have relatively long spines. Anadromous completely plated populations tend to have longer spines than freshwater populations, which may reflect a greater risk of predation in the sea, although there is no quantitative evidence of differential predation rates in the two environments (Hagen and Gilbertson, 1972; Gross, 1978). In Europe, the extent of serration of the spines and the size of the pelvic skeleton were also correlated with the distribution of fish predators (Gross, 1978).

In encounters with invertebrate predators which grasp the stickleback and do not attempt to swallow it whole, spines may be a disadvantage because they may make it easier for the predator to maintain a grip on the prey (Reimchen, 1980; Reist, 1980b).

The role of the lateral plates as a defence against predation is not clear. Experimental evidence shows that the anterior lateral plates can provide important structural support for the dorsal and pelvic spines when they are erect (Reimchen, 1983). The plates may also help to prevent damage to the stickleback if it is captured by a predator and so increase its chances of escape and subsequent survival. Under some circumstances, predation does seem to exert a selective effect on the number of lateral plates (Chapter 10).

The effectiveness of these morphological adaptations against predation varies with the type of predator and its size. Those predators, perch and trout, which manipulate the stickleback in their mouth while attempting to swallow it, and also lack the specialised dentition of a piscivore, tend to let the stickleback escape more often than does the pike (Moodie *et al.*, 1973). In a few populations, *G. aculeatus* may reduce predation by achieving an unusually large body size, so that only large and relatively rare predators can successfully take them (Moodie, 1972a,b) (Chapter 10). In other populations, the same species may evade predation by becoming sexually mature at a size below that which is profitable for the predator to take (McPhail, 1977). Until there are critical experimental studies on the size selection by the usual predators of sticklebacks, these suggestions remain only plausible hypotheses. The role of bird predation in selecting for morphological defence mechanisms in the stickleback has been neglected but is probably as significant as fish predation (Reimchen, 1983).

Cryptic Coloration

An important primary defence of the sticklebacks is their coloration. Typically, the dorsal surface is a drab olive or brown with patches of darker brown. The ventral surface is paler and may be silverish. Against a substrate or in vegetation, this colour scheme makes the stickleback visually inconspicuous. They also have the ability to adapt their coloration to the substrate. When living among algae, *S. spinachia* tends to an olive green, but brownish when over sandy or muddy substrates, and will adjust its colour according to the environment. When *G. aculeatus* is transferred from a dark to a pale substrate, the dorsal surface pales within a minute whereas the dark lateral stripes pale over a 5-min period. When the fish is placed back on a dark surface, the stripes darken within a minute and the dorsal surface within 5 min (Burton, 1975, 1978). *Culaea inconstans* shows colour changes that make it more cryptic when it first detects the presence of a pike (Reist, 1983).

During the breeding season, males become more conspicuous and in several species adopt specific breeding colours, but colours that make a courting male obvious to a female may also make him obvious to a predator. In *G. aculeatus*, which breeds in relatively open areas and also has the most flamboyant breeding colours, there is evidence that the extent to which the breeding colours develop can be related to the intensity of heterospecific predation.

In Lake Wapato and in Meyer Lake, both in north-west America, populations were found in which only a small proportion of the breeding males developed the normal red throats during the breeding season. Both populations experienced predation by trout, and there was evidence that the failure of most of the males to develop the normal breeding colours was a consequence of this predation (Chapter 10). Given the wide area of sympatry between *G. aculeatus* and its major predators, there are still relatively few reports of populations comparable to those in Lake Wapato and Meyer Lake, so the circumstances in which a proportion of the males will fail to develop their typical breeding colours are still obscure.

Behavioural Responses to Predators

Behaviour in relation to predators can form either a primary or a secondary defence. The stickleback may choose to spend its time in habitats that make it less likely to be detected by a predator, a primary defence which is also correlated with the possession of a cryptic coloration. Once the stickleback has detected the presence of a predator, it can display a repertoire of behaviour that will reduce its risk of capture. Although the species so far studied have relatively similar repertoires, the balance between the primary and secondary behavioural defences varies.

In comparison with *P. pungitius* and *C. inconstans*, *G. aculeatus* seems to rely less on a primary defence and more on a behavioural secondary defence and its morphological adaptations. When the responses of *P. pungitius* and *G. aculeatus* placed in a strange aquarium were compared, *P. pungitius* was more likely to enter weeds, and having entered spent longer in them. When exploring the tank, *G. aculeatus* tended to move smoothly throughout the tank whereas *P. pungitius* moved rapidly between patches of weed (Benzie, 1965). *Culaea incon-stans* is a shy fish usually associated with thick vegetation. Correlated with these differences, there are the variations in the choice of nesting sites described in Chapter 7. By virtue of its longer spines and better developed armour, *G. aculeatus* has become relatively emancipated from thick vegetation and builds its nest on the substrate, frequently in open situations.

An undisturbed stickleback moves through its environment by sculling with its pectoral fins, a mode of locomotion that produces jerky swimming. If it detects a potential predator, the stickleback stops and tends to fix the potential danger with binocular vision. *Pungitius pungitius* makes its initial reaction at a greater distance from the predator than *G. aculeatus*. The stickleback may then approach. At this stage, *G. aculeatus* is more likely to approach than *P. pungitius.* The mode of swimming may also change to a smoother locomotion using the body and tail. A variety of 'escape' responses may be shown. At an early stage in the encounter the stickleback may sink towards the bottom. It may then back away using pectoral sculling or swim smoothly out of danger. Two extreme escape responses are jumping, a rapid movement in an unpredictable direction away from the pre-dator, and freezing, in which the stickleback becomes completely motionless apart from slight opercular movements. The spines are frequently erected during the encounter. *Pungitius pungitius* and *C. inconstans* are more likely to conceal themselves in vegetation than *G. aculeatus*, sometimes wedging themselves in the vegetation. When concealed in weed *A. quadracus* and *P. pungitius* will take up a head-down vertical position in which they freeze. When frightened, *C. inconstans* and *A. quadracus* will also attempt to bury themselves partially in the substrate (Wootton, 1976).

A detailed analysis of the behaviour of *G. aculeatus* in the presence of a pike suggested that the sticklebacks' behaviour could be described in terms of a 'precaution-investigation' factor and a 'boldness-timidity' factor. The first factor contrasts behaviour that tends to maintain a distance between stickleback and pike with behaviour that tends to bring the stickleback close to the pike, allowing an investigation of the predator. The second factor contrasts the behaviour of an undisturbed fish showing jerky swimming and a tendency to approach the pike with that of a fish whose lack of jerky swimming and ten-dency to keep the dorsal spines erect suggests a timidity in the presence of the pike (Huntingford, 1976a). The behaviour of *C. inconstans* changed significantly when it detected the presence of a pike. The time spent in jerky swimming decreased, but the times spent frozen, lodged in vegetation, and with dorsal spines erect, all increased (Reist, 1983).

Prior experience of a predator influences the subsequent behavioural response of the stickleback. *Gasterosteus aculeatus* which had had prior experience of a stalking pike responded to a stationary pike at an average distance of 390 mm compared with 190 mm when naive. The experienced fish were more likely to retreat into weed, and were less likely to approach the pike before taking fright (Benzie, 1965). There was some evidence that the amount of parental care a stickleback had experienced after hatching affected its subsequent response to a pike, although this effect was neutralised by experience with a predator. The young of *G. aculeatus* whose father had been removed at hatching were less likely to retreat into weed and more likely to approach a pike than normally fathered fish. When the fry were kept with their father for an unusually long period, 6 weeks, they subsequently responded to a pike at a shorter distance but were less likely to approach it than normal fish (Benzie, 1965). It is difficult to judge whether these effects have any significance in natural populations, given the lack of information on the normal length of time for which fry and father remain in proximity (Chapter 7).

A comparison of *G. aculeatus* with *P. pungitius* shows that the former is bolder in the presence of a potential predator than the latter, so the behaviour of *G. aculeatus* puts it at greater risk. It also tends to be more aggressive than *P. pungitius*, a difference correlated with the difference in typical nesting sites (Benzie, 1965; Wilz, 1971). This raises the possibility of a relationship between behaviour towards a predator and aggressive behaviour in sticklebacks. The tendency for male sticklebacks to be most aggressive towards both conspecific and heterospecific territorial intruders once there are fry in the nest was described in Chapter 7.

Male *G. aculeatus* show individual differences in the level of their aggression, and these differences are consistent over a breeding cycle. Males that are most aggressive when they have an empty nest also tend to be the most aggressive when they have fry in their nest. Additionally, males that are most aggressive to a conspecific intruder tend to be most aggressive to a heterospecific intruder (Huntingford, 1976a). A comparison of the behaviour of males towards a pike outside the breeding season with the behaviour of the same males towards a conspecific male during the breeding season showed that there was a positive correlation between the scores of the males on the 'boldness-timidity' factor from the pike tests with their scores on the 'aggression' factor from the conspecific tests. Males that were highly aggressive towards conspecifics in the breeding season tended to be boldest in the presence of a pike outside the breeding season (Huntingford, 1976a). A possible explanation for this covariation between conspecific aggression and the response to a predator is that it allows an economic and efficient way of regulating the levels of these two behaviour patterns relative to the predatory pressure in the environment. During the sexual cycle, male conspecifics are rivals for nesting sites and females but are also important predators of eggs. A highly aggressive male with well developed intimidatory breeding colours is likely to be more successful than a less aggressive or drabber male. But the aggressive behaviour and coloration will probably make the male more susceptible to heterospecific predation, so that in environments where this risk is high, the fish should be less aggressive, less highly coloured and more timid of predators (Huntingford, 1976a). Some support for this hypothesis was obtained by comparing the behaviour of male sticklebacks from several populations in Scotland. The response of males to a pike and to a breeding male were measured. Males that came from populations that were judged to suffer from no or light predation tended to score significantly higher on the 'boldness' factor than those from populations judged to experience high predation pressures. There were also significant inter-population differences in the mean aggression shown to the conspecific. There was a significant correlation between the 'boldness' and aggression scores, that is, populations with high scores for conspecific aggression also tended to have high scores for 'boldness' towards a predator. There was a positive but non-significant correlation between 'boldness' and aggression to conspecific females (Huntingford, 1982).

Most of the studies on the behaviour of sticklebacks in the presence of predators have concentrated on fish predators. Sticklebacks show similar patterns of behaviour if alarmed by a potential bird predator (Wootton, 1976; Giles, 1984), though the responses they show may depend on the extent of their experience of fish and bird predation.

If the sticklebacks are in a school when alarmed, the school becomes denser as the fish move closer together and they swim rapidly away from the source of alarm (Keenleyside, 1955). Schooling behaviour may reduce the chances of an individual stickleback being caught by a predator (Chapter 4) (Bertram, 1978).

Inter-population and inter-specific differences in the behaviour towards predators and other correlated behaviours provides considerable scope for the analysis of the role of predation as a selective factor in populations (Chapter 10).

Quantitative Aspects of Predation

The least well studied aspect of predation is the estimation of the rate at which stickleback populations suffer losses from predation and the extent to which these losses vary with the age and size of the sticklebacks. The estimates of predation by loons living on Boulton Lake described earlier indicate that a single pair of birds rearing young may take over 100 000 *G. aculeatus* from a population. Correspondingly high rates of predation by breeding terns in Finland were suggested by the estimates of Lemmetyinen (1973). No similar estimates are available for the potential rates of loss through fish predation. In Lake Wapato, during winter 80 per cent of the rainbow trout with food in their stomachs had eaten *G. aculeatus*, whereas in spring only 10-12 per cent had eaten sticklebacks. The average number of sticklebacks per stomach was 4.8 in winter and 2.4-3.0 in spring (Hagen and Gilbertson, 1973a). The rate of food consumption by trout is low in winter because of the low water temperatures, which mean low rates of digestion (Elliott, 1979). So, although a high proportion of trout were eating sticklebacks, this does not necessarily indicate a high rate of exploitation. Until there are accurate estimates of the rate of food consumption by the predators of sticklebacks, together with reliable estimates of the density of the predators, the effect of predation on the abundance of stickleback populations will remain unknown.

Parasitism

Although the effect of parasitism on the victim is less immediately dramatic than that of predation, a parasitised stickleback can suffer from a syndrome of deleterious effects. Sticklebacks are hosts to a wide range of parasites (Wootton, 1976), but the effects on the host of only a few of these are known.

The best studied of these parasites and the one that has the most obviously spectacular effects is a cestode, *Schistocephalus solidus*. This tapeworm has a complex life history in which the stickleback plays a vital role. The adult worm reaches sexual maturity in the alimentary canal of a fish-eating bird. Eggs pass out in the faeces of the bird and if they fall into water they hatch into free-swimming coracidia, which infect copepods. In the copepod the larvae enter the procercoid stage. If an infected copepod is eaten by a stickleback, the larvae migrate to the body cavity of the fish where they start the plerocercoid stage. It is in the stickleback's body cavity that the main growth of the worm takes place. This growth can be so spectacular that a stickleback ends up carrying a weight of *S. solidus* greater than its own body weight. Such a stickleback has a grossly distorted body with a massively swollen abdomen (Figure 8.4). The parasite's life history is completed when an infected stickleback is predated by a bird, for the cestode becomes sexually mature within about 36 h of ingestion. Its sojourn in the bird is usually only a few days, which may represent an adaptation to the migratory habits of many of its bird hosts (McPhail and Peacock, 1983). A rapid maturation will ensure that eggs have a good chance of being deposited in the body of water from which the infected stickleback came.

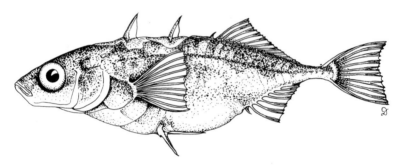

Figure 8.4: Abdominal Distension of *G. aculeatus* Caused by *Schistocephalus solidus*. Fish Collected in Llyn Frongoch, Wales

Transplantation experiments have suggested that *S. solidus* from *G. aculeatus* cannot live in *P. pungitius* (Orr *et al.*, 1969), although the parasite has been recorded from both *P. pungitius* and *C. inconstans*. It is possible that there are several species of *Schistocephalus*, each specific to a different stickleback (McPhail and Peacock, 1983).

Another parasite for which the stickleback is an intermediate host, and a fish-eating bird the final host, is the digenetic trematode *Diplostomum*. One species, *D. gasterostei*, lives as a metacercaria in the retinal layer of the stickleback's eye, whereas metacercariae of *D. spathaceum* occur in the lens. Sticklebacks are infected by cercariae of *Diplostomum* released by infected water snails.

Sticklebacks are also the final hosts of several parasites such as the acanthocephalan *Acanthocephalus clavula* and the cestode *Proteocephalus filicollis*, which live and come to sexual maturity in the stickleback's intestine.

Effects of Parasitism

Metabolism and Growth. Schistocephalus solidus has a significant effect on the energy partitioning of its host *G. aculeatus* (Chapter 5). Because of the growth rate and size that the plerocercoids can achieve, they make significant demands on the supply of energy and materials of the host. Little is known about the environmental conditions in the body cavity of the stickleback so that it is unclear to what extent the *S. solidus* is metabolising aerobically or anaerobically. In inactive fish, there were relatively slight differences in the rate of respiration of parasitised and uninfected fish of a similar length. When active, the parasitised fish had a higher rate of respiration, which may reflect, at least partly, the extra cost of locomotion of fish with swollen abdomens (Lester, 1971; Meakins and Walkey, 1975). The rate of respiration of parasitised fish was always greater than that of unparasitised fish of the same weight when the comparison excluded the weight of the parasite. If the weight of the parasite were included in the total weight of parasitised fish, no consistent effect of parasitism was found (Meakins and Walkey, 1975).

Parasitised *G. aculeatus* had slightly greater gross growth efficiencies than unparasitised fish when fed *Tubifex* ad lib: 7.5 per cent and 6.7 per cent, respectively. But in terms of growth of fish flesh as distinct from the growth of fish and parasite flesh added together, the parasitised fish were far less efficient, with a gross growth efficiency of only 2.7 per cent (Chapter 6) (Walkey and Meakins, 1970; Meakins, 1974a). So, although the parasitised fish grows more efficiently, the beneficiary is the *S. solidus* rather than the stickleback.

This effect of *S. solidus* on growth was detected in a natural population of *G. aculeatus* in Priddy Pool (England) (Pennycuik, 1971a). At 2 years old, the parasitised sticklebacks were estimated to be about 25 per cent lighter than uninfected fish of the same age. In this population both *S. solidus* and *D. gasterostei* had a negative effect on the condition of the sticklebacks, but the effect of the cestode was far greater than that of the trematode.

Both field and experimental studies have shown that sticklebacks carrying *S. solidus* have significantly smaller livers than uninfected fish and they usually lack the visceral fat bodies (Arme and Owen, 1967; J.R.M. Allen and R.J. Wootton, unpublished).

Reproduction. The fecundity of female sticklebacks is a function of their size (Chapter 7), so the inhibition of growth by *S. solidus* will reduce the reproductive potential of infected females. There is also evidence that the cestode has more direct effects on reproduction, particularly in females. Egg production by the females makes heavy demands on their energy income (Chapter 7), so energy and materials which are diverted by the cestode are not available to meet the costs of egg production. In the experimental studies on the effect of food ration on the fecundity of females, described in Chapter 7, females that were infected by *S. solidus* sufficiently to cause abdominal distension would not spawn although they were courted vigorously by the males entranced by their magnificent bellies (R.J. Wootton, unpublished). Collections of females from natural populations suggest that vitellogenesis is delayed in heavily infected females, whose ovaries do not reach the size characteristic of spawning females (Arme and Owen, 1967; Meakins, 1974b). After the spawning season, the infected females had relatively larger ovaries than uninfected fish, and these ovaries contained high proportions of pre-ovulatory corpora atretica. This indicates that infected females are less likely to mature eggs to the stage at which they can be spawned. There is no evidence that the cestode disrupts the hormonal control of the female stickleback's reproduction; the main effects seem to be a consequence of the energy demands the parasite is making. In male sticklebacks, infection with *S. solidus* does not interfere with spermatogenesis and infected males will build nests, though if the males are heavily infected, the nests may not be completed (Arme and Owen, 1967). There is no evidence on whether infection interferes with the development of the breeding colours of the males or affects the mate choice by courting females.

Mortality. Heavy infections by a parasite either alone or in combination with other parasites may directly cause the death of the stickleback host. Infection by parasites can also increase the risk of death from other causes, so although the parasite is not directly causing the death of its host, the death rate among the parasitised fish is higher than in the non-parasitised sticklebacks.

The ciliate fish parasite *Ichthyophthirius*, the dreaded 'white-spot' feared by all aquarists, was responsible for the deaths of almost the complete 0^+ year class of *G. aculeatus* during a heavy infection in a loch in Scotland (Hopkins, 1959). In a Newfoundland population of *G. aculeatus*, mass deaths were associated with heavy infections of *S. solidus* and the fish louse, *Argulus canadensis*, an ectoparasite (Threlfall, 1968). Dead sticklebacks collected in Priddy Pool in late autumn had, on average, a higher number and higher mean weight of *S. solidus* than live fish at the same time (Pennycuik, 1971a,b,c). *Culaea incon-stans* that were heavily infected with the metacercariae of the digenean gut parasite *Apatemon gracilis pelludicus* were inferred to have a higher death rate than more lightly infected fish (Gordan and Rau, 1982).

For those parasites such as *S. solidus* and *D. gasterostei* for which the stickleback is an intermediate not a final host, the death of the fish is also their death unless it results in their transmission to the final host. Such a transmission is more likely if the parasite causes changes in its host that make the latter more likely to be predated by the final host, typically a bird. *Gasterosteus aculeatus* carrying *S. solidus* become bloated, and hence visually more conspicuous. When heavily infected, the stickleback tends to move slowly near the water surface. Although they are capable of rapid escape movements over short distances, their respiratory costs during fast swimming are significantly higher than those of uninfected fish (Lester, 1971). The changes in locomotor ability, together with the tendency to move close to the water surface, predispose the infected fish to predation by birds. Fish parasitised with *S. solidus* recovered faster than uninfected fish after they had been exposed to a frightening overhead stimulus, a response which may also make the parasitised fish more susceptible to bird predation (Giles, 1983b). Sticklebacks heavily infested with metacercariae of *Diplos-tomum*, which lodge in the eye, may have poorer vision than infected fish and so be less likely to detect the approach of a predator. Field evidence that parasitised and unparasitised sticklebacks experience differential rates of bird predation is still required. In an Alaskan lake, differential predation by fish on *G. aculeatus* parasitised with *S. solidus* could not be demonstrated (Gilbertson, cited in McPhail and Peacock, 1983).

When subjected to starvation, *G. aculeatus* carrying *S. solidus* survive for a significantly shorter period than uninfected fish (Walkey and Meakins, 1970; Pascoe and Mattey, 1977; Pascoe and Woodworth, 1980). Pennycuik (1971a,b) suggested that parasitised sticklebacks in Priddy Pool suffered higher mortality rates during periods of food shortage. Field estimates of the rates of food consumption in *G. acu-leatus* show that late autumn and winter can be a period of very low rates of feeding (Chapter 4). *Gasterosteus aculeatus* infected with *S. solidus* had shorter median periods of survival in toxic solutions of cadmium than uninfected fish (Pascoe and Cram, 1977). When the sticklebacks were maintained on a ration of enchytraeid worms well below that required for them to maintain a constant body weight, parasitised fish in a dilute solution of cadmium ($0.32\,\mathrm{mg\,litre^{-1}}$) had a significantly shorter median survival time than unparasitised fish. These results suggest that parasitised sticklebacks will suffer higher rates of mortality than uninfected fish when the quality of the environment deteriorates, and that the parasitised fish are particularly vulnerable if several factors that cause environmental deterioration are present simultaneously.

Stickleback-Parasite Relationship

Compared with the predator-prey interaction, the parasite-host relationship is ambiguous. If the parasite plays a role in causing the death of the host at an inappropriate time or place, the parasite as well as the host dies. Yet the parasite must exploit the host at a sufficiently high rate to maximise its own rate of reproduction. Temporal changes in both the proportion of the host population infected with a parasite and the mean number and variance of the parasites per host suggest how the parasite and host coexist. Typically, the distribution of the parasite in a population of sticklebacks is over-dispersed (Pennycuik, 1971b; Gordon and Rau, 1982). If a given parasite were distributed at random in a stickleback population, the numbers of parasite per fish would follow a Poisson distribution in which the mean and variance are equal. In practice, the variance is usually greater than the mean, indicating an over-dispersed distribution. The negative binomial, logarithmic or log-normal functions can be used to describe over-dispersion (Pennycuik, 1971b). Its biological significance is that a relatively few fish in the population carry a relatively high number of parasites, and even with high mean parasite abundance a relatively high proportion of the host population has a low or zero infection. Fish with a high number of parasites may be more effective at transmitting the parasite to its final host than lightly infected hosts, so over-dispersion may enhance the transmission of the parasite.

Three parasites of the population of *G. aculeatus* in Priddy Pool had over-dispersed frequency distributions. The distributions of the acanthocephalan *A. (Echinorhynchus) clavula* and *D. gasterostei* were such that the commonest class of sticklebacks were those carrying no parasites, even when the mean number of parasites per fish was relatively high. In contrast, a high proportion of the population was carrying *S. solidus* and the commonest fish were those carrying one or two parasites. For this parasite, it was its weight rather than its number per fish that caused damage to the host. Few fish were collected in which the combined weight of fish and parasite exceeded 3 g or the *S. solidus* had a weight greater than that of the host. On average about eleven plerocercoids had a mean weight equal to that of the mean weight of the fish. In the total sample collected, 8.4 per cent of the fish carried more than eleven plerocercoids, but these heavily infected fish were carrying 40.9 per cent of all the *S. solidus* in the sample. So, even with the high percentage of infected sticklebacks, a high proportion of the parasitic burden was carried by a small proportion of the population. These heavily parasitised fish would be particularly susceptible to predation because of the distortions caused by the plerocercoids (Pennycuik, 1971a,b,c).

In the Priddy population, the percentage of fish infected and the mean number of parasites per fish were functions of the age and length of the sticklebacks. In fish less than 25mm in length, infection with *S. solidus* was rare. There was a rapid increase in the percentage of fish infected as their size increased, over 90 per cent of fish 40mm and longer being infected, with the exception of the longest fish which had about a 70 per cent rate of infection. The mean number of parasites per fish also increased with fish size, reaching a maximum in fish between 60 and 65mm in length. Longer fish had a significantly lower mean number of *S. solidus*. The decline among the largest fish in both the percentage infected and the mean number of plerocercoids per fish suggests that the heavily infected large fish were dying before the more lightly infected sticklebacks (Pennycuik, 1971d). A similar pattern of infection with *S. solidus* was found in a population of *G. aculeatus* in Boulton Lake on the Queen Charlotte Islands, although in this population the proportion of fish infected was lower. The Boulton Lake population also carried the cestode *Cyathocephalus truncatus*, the sticklebacks becoming infected by eating amphipods (Reimchen, 1982). Only fish greater than 30mm in length were infected with *C. truncatus*, probably because smaller fish were unable to consume the amphipods whereas they could eat copepods, thus becoming infected with *S. solidus*.

The Priddy population showed a steady increase with length in both the percentage of fish infected with *D. gasterostei* and the mean number per fish. Both indices reached a maximum in the largest fish, although there was some evidence that the most heavily infected fish in the older age classes were dying. The percentage of the populations infected with *A. clavula* increased rapidly in fish greater than 30mm in length, reaching a maximum for fish between 60 and 70mm but then dropping sharply. The mean number of worms per fish showed a similar pattern. *Acanthocephalus clavula* differs from *S. solidus* and *D. gasterostei* in that it reaches sexual maturity in the stickleback, dies and is voided, whereas there is no evidence that the other two species suffer significant mortality in the host so infection is cumulative (Pennycuik, 1971d). Significantly, the parasite for which the stickleback is the final host had no detectable effects on the growth or condition of the sticklebacks (Pennycuik, 1971a).

Superimposed on the changes with body size there were also seasonal changes in the Priddy population. The infections by all three parasites increased in summer and early autumn and decreased in winter and spring. For *S. solidus* there was a marked increase in the mean number and weight of plerocercoids per fish in the period between July and October (Pennycuik, 1971b). An almost identical pattern in the percentage of *G. aculeatus* infected with *S. solidus* occurred in Fuller Lake on Vancouver Island. In this population, there was an inverse relationship between the percentage of gravid females in the population and the mean weight of the parasite as a proportion of the total fish weight (McPhail and Peacock, 1983). In Boulton Lake, the percentage of fish infected with *S. solidus* that were captured in minnow traps increased sharply between June and July, though a higher percentage of infected fish were caught in seine nets in the same region (Reimchen, 1982). One interpretation of this pattern of a rapid increase in infection in late summer is that the parasite-host interaction has evolved so that the parasite does not exert its major deleterious effects on the host population until after the peak reproductive period for the fish (McPhail and Peacock, 1983). The temporal pattern of reproduction and infection in several populations needs to be determined accurately before this suggestion can be evaluated. Because of the mode of infection with *S. solidus*, the pattern of infection will depend on the temporal sequence of the appearance of the fish-eating birds, the copepod initial host and the use of the copepods as food by the sticklebacks. Late summer is a period when copepods are often an important component in the diet of sticklebacks and high water temperatures will permit high rates of food consumption, so that the conditions are favourable for the successful transmission of the parasites.

Effect of Parasitism on Stickleback Abundance

The role of parasites in determining the population density of their hosts has attracted considerable theoretical analysis recently (Anderson, 1981). In the stickleback it is likely that parasites such as *S. solidus* can increase the death rate in a population both directly and by making the fish more susceptible to predation, starvation and pollutants while also decreasing the birth rate by reducing the proportion of females that become gravid. On the basis of her study of the Priddy Pool population, Pennycuik (1971a) speculated that infection with *S. solidus* caused cyclic fluctuations in population size. She argued that the infection builds up until a high proportion of the fish are carrying plerocercoids, some with heavy infections. This leads to a high death rate and an inhibition of breeding so the fish population declines and with it the size of the parasite population. This reduction reduces the chances of successful transmission to the final host so the infection in the fish population becomes lighter. As a consequence, reproduction becomes more successful and the stickleback starts to increase again, which allows the *S. solidus* population to increase in turn and the cycle restarts. As Chapter 9 shows, knowledge of the population dynamics of stickleback populations is too scanty to assess whether this is a realistic model of the effect of *S. solidus* on abundance.

Inter-specific Competition

When populations of two or more species are living sympatrically, they may have overlapping resource requirements. If those resources are in short supply, then the populations will be in competition for what resources are available. This competition may take the form of direct confrontations between individuals from the populations, typically in the form of inter-specific aggression. Alternatively, the competition may not involve any direct confrontations but result because a resource that has been exploited by one of the populations is not then available to the others.

Competition and predation are usually considered the major interactions between species that can determine the number of species that coexist in the same area (Pianka, 1981). However, direct evidence for competition in natural habitats has proved elusive. The demonstration of competition requires that the presence of a potentially competing species has one or more of the following effects on a population: an increase in the death rate, a decrease in the birth rate, an increase in emigration, a decrease in immigration, a decrease in the average growth in size of individuals or a decrease in the area inhabited. If competition is or has been important for a species, then it is assumed that natural selection will favour characters that reduce the intensity of competition by minimising the overlap in requirements for resources such as food and space.

Competition should be most intense between closely related species that have similar ecological requirements and occur sympatrically. The sticklebacks are such a group. In many areas *G. aculeatus* and *P. pungitius* occur sympatrically as do *P. pungitius* and *C. inconstans*. On the east coast of North America between Newfoundland and New Jersey, up to four species can occur sympatrically (Chapter 2). These situations should provide clear examples of the process of competition and its consequences.

Competition Between Sticklebacks

Competition for Food. Populations of *G. aculeatus* and *P. pungitius* in the River Birket in north-west England had almost identical diets, although *P. pungitius* was less likely to eat algae than was *G. aculeatus*. There was no direct evidence that this dietary overlap led to competition between the species, though both had relatively slow growth rates (Hynes, 1950). In contrast, the same two species living in Matemak Lake in Quebec had distinctly different diets, *G. aculeatus* feeding primarily on zooplankton whereas *P. pungitius* fed on cladocerans and chironomids from the littoral zones of the lake (Coad and Power, 1973a).

Three species, *G. aculeatus*, *G. wheatlandi* and *P. pungitius* were found breeding in salt-marsh pools in the Isle Verte region of the St. Lawrence Estuary (Quebec), occupying the pools from late April until the end of the breeding season. Stickleback densities were high, up to $20\,\mathrm{m}^{-2}$, but no other fish species were present. The pools were flooded about twice a month by the tide so there could be changes in the population densities of the three species through immigration and emigration. There was considerable overlap in the diets of the species, with stickleback eggs, chironomid larvae and pupae, copepods, gammarids and corixids forming the major food items. Some inter-specific differences existed in the relative proportions of these items in the diet. In May and June, *G. wheatlandi* ate a higher proportion of copepods, *G. aculeatus* fed more on eggs, gammarids and chironomid larvae, and *P. pungitius* fed most heavily on chironomid larvae and corixids. These differences tended to disappear in July at the end of the breeding season when all species fed on gammarids and chironomid pupae. In early summer the average size of the food items eaten by *G. aculeatus* was bigger than for *P. pungitius*, which was eating significantly larger items than *G. wheatlandi*. These differences had also largely disappeared by July, although *G. wheatlandi* still tended to eat smaller items than the two other species. In all species, the average size of prey tended to increase over the summer. Experiments in which different combinations of the species were held in enclosures in the pools did not provide any evidence that the presence of one species affected the diet of another (Worgan and FitzGerald, 1981b; G. Walsh and G.J. FitzGerald, pers. comm.).

The average size of the adults in the pools differed. In terms of length, *G. aculeatus* was 1.4-1.5 times bigger than *P. pungitius* and the latter was 1.3 times longer than *G. wheatlandi*. Mouth widths also differed by similar ratios. *Apeltes quadracus*, which was present in the tidal creeks of this area but absent from the pools, was almost identical in size to *P. pungitius*. Some ecological thoery has suggested that ratios of these magnitudes are characteristic of coexisting, closely related species that have evolved to minimise competition and so allow the coexistence (Pianka, 1981). There was no direct evidence that competition for food was occurring in the pools, but the use of enclosure experiments in which the species composition can be manipulated and controlled will permit studies on birth, death and growth rates in the three species.

Competition for Space. During the breeding season, male sticklebacks are strongly territorial (Chapter 7) and there is a maximum number of males that can be accommodated in a given area. Aggressive defence of the territory is directed against both conspecifics and heterospecifics, so when sticklebacks are sympatric there is the possibility of direct, interactive competition between the males for breeding territories.

Laboratory experiments showed that *G. aculeatus* males displaced territorial male *G. wheatlandi* or *A. quadracus* even when the latter species had established territories and built nests before the introduction of *G. aculeatus*. This species was significantly more aggressive to a conspecific male than either *G. wheatlandi* or *A. quadracus* and significantly more aggressive to a heterospecific male than *A. quadracus* (Rowland, 1983a,b). In these experiments, the male *G. aculeatus* were larger than males of the other two species. These results suggest that, in competition for territorial space, *G. aculeatus* males will, because they are more aggressive, outcompete both *G. wheatlandi* and *A. quadracus*.

In the St. Lawrence salt-marsh pools, three species did breed within the same pools at the same time, although *P. pungitius* was largely restricted to nesting in thick algae and did not frequent the open areas of the pools once the algae stands had grown. Both the *Gasterosteus* species nested in open areas, in some cases adjacent to each other. Most of the aggressive behaviour of male *G. aculeatus* was directed at conspecifics, whereas that of *G. wheatlandi* and *P. pungitius* was directed at conspecifics and heterospecifics in approximately the proportions in which the species occurred (FitzGerald, 1983). It is not yet clear whether there was competition for space between the males. This would involve showing that there were males present that were physiologically capable of establishing a territory and building a nest but were failing to do so because of the presence of males of the other two species. It is possible that competition between conspecific males (Chapter 7) overrides any effects of competition between heterospecific males for territories. A further complication is the regular flooding of the pools. At any one time, the species composition in a pool may merely reflect the number of sexually mature males that were available to colonise that pool during a flood.

Gasterosteus and *A. quadracus* do coexist in salt marshes in other areas of their sympatry, but the tendency for *A. quadracus* to nest in vegetation may minimise competition between them for territories (Rowland, 1983b). It may be significant that *P. pungitius* is present in the St. Lawrence system, because this species is similar in size to *A. quadracus* and also nests in vegetation. It may therefore be the presence of *P. pungitius* rather than the *Gasterosteus* species that excludes *A. quadracus* from the pools although it does breed in the adjacent tidal creeks (Worgan and FitzGerald, 1981a).

Female sticklebacks in the St. Lawrence pools also showed differences in their spatial distribution. Female *P. pungitius* spent most of their time in the algae rarely venturing into open areas. When feeding, *G. aculeatus* females spent a relatively high proportion of their time near the bottom whereas *G. wheatlandi* spent more time near the surface or in algae (G. Walsh and G.J. FitzGerald, pers. comm.).

Timing of Breeding in Sympatry. One way in which stickleback species could minimise competition for breeding territories would be to breed at different times in the year. However, there is a restricted period in the year in which sticklebacks can reproduce successfully so total separation of the breeding periods is not possible. Even when the species are sympatric there is considerable overlap in their breeding periods, but there are some differences that may reduce any interspecific competition (Chapter 7).

Conclusion. The coexistence of four species of sticklebacks on the east coast of North America provides a valuable opportunity for both field and experimental studies on the role, if any, of competition between closely related sympatric species. A major factor mediating any competition during the reproductive season is the extent to which the males prefer to nest in or closely associated with vegetation. This preference can also be correlated with differences in the reproductive biology and response to predators of the sticklebacks (Chapter 7, and above). As yet, there is a lack of evidence that competition has an important effect on the rates that determine the changes in abundance of stickleback populations.

Competition with Salmonids and Other Fish Species

The catholic diet of the sticklebacks (Chapter 4) means that they frequently take prey that are taken by other teleosts. Of particular interest is the possible interaction between *G. aculeatus* and the young stages of the migratory Pacific salmon (*Oncorhynchus* spp.). The biology of salmon is intensively studied because of their economic importance to the countries that border the north-west Pacific.

Many of the lakes that act as nursery grounds for the sockeye salmon (*O. nerka*) also contain populations of *G. aculeatus* and *P. pungitius*, which are similar in size to the sockeye fry. In the Wood River lakes of Alaska, the diet of the sockeye fry and sticklebacks showed considerable overlap. When pelagic, both sockeye and *G. aculeatus* preyed on zooplankton though there were some differences in emphasis, with the sockeye taking more cladocerans and the sticklebacks more copepods. The sockeye also took winged insects off the surface, which the sticklebacks rarely did. When they were in the littoral zone, both species ate cladocerans, copepods, and chironomid larvae and pupae. Sticklebacks relied more on benthic food items than the sockeye, which readily took food from the surface (Rogers, 1968). There were also significant correlations between the diet of *G. aculeatus* and young sockeye during spring and summer in Great Central Lake, Vancouver Island (Manzer, 1976). Lake Dalneye in the Kamchatka region of the USSR contains populations of *G. aculeatus* and sockeye fry, both of which feed predominantly on zooplankton (Krokhin, 1970). These dietary overlaps suggest that the two species might be competing for food, but evidence of direct competition is equivocal.

During a 30-year period, the population of sockeye in Lake Dalneye has gone through three phases. Initially the population was high, then, in the 1940s, it declined after the failure of the 1943 and 1944 generations. During the third phase, from the mid-1950s, an intensive highseas fishery for sockeye kept the population at low levels. The abundance of sticklebacks increased during the second phase and at their peak it was estimated that they were consuming about a third of the total consumption of zooplankton by fish in the lake. The third phase was marked by a decline in the abundance of stickle-backs as the lake became less fertile because there was a reduced input of nutrients into the lake from the decaying corpses of the spawned-out sockeye. The increase in stickleback abundance during the second phase, which coincided with a decline in the abundance of sockeye fry, does suggest that the sticklebacks were benefiting from the sockeye decline but direct evidence for this is lacking (Krogius *et al.*, 1972; Krogius, 1973). In a lake in the Wood River system, there was an inverse correlation between the mean size the sticklebacks had reached at 1 year old and the abundance of sockeye fry in the previous year, which again suggests that the presence of sockeye fry has an adverse effect on the sticklebacks. There was also an inverse correlation be-tween the mean catch of sockeye fry and the mean lengths of all the fish species in the lake, which suggests that food supply sets a limit to the total fish biomass the lake could support (Rogers, 1973). Again, direct evidence of competition between sticklebacks and sockeye is lacking.

The clearest evidence for the adverse effect on *G. aculeatus* of the presence of a competing fish species comes from the Lake Wapato study (Chapters 7 and 10). Sticklebacks in this lake usually live for just over a year so heavy losses in a single generation have catastrophic effects on population abundance. In 1969, the lake had a large popu-lation of sticklebacks that were preyed on by rainbow trout and a small population of the pumpkinseed sunfish, *Lepomis gibbosus*. A collection of young-of-the-year fish (YOY) made in September after the end of the breeding seasons of sticklebacks and sunfish yielded average seine hauls of 134 sticklebacks and 4.5 sunfish. A year later, sticklebacks were observed breeding successfully, but there was also an eightfold increase in the number of sunfish nesting in the same area as the sticklebacks. The collection in September yielded only 1.7 sticklebacks per seine but 88.6 sunfish. There was no recovery in the stickleback population over the next 5 years. Although the mechanism of competition between sticklebacks and sunfish is not known, it is likely that the YOY sunfish outcompeted the stickle-backs in the areas of the lake where the two species were breeding (Kynard, 1979b).

Conclusion

Ecological theory assigns major roles to the processes of predation, parasitism and competition in determining the abundances of animal populations and of species that can coexist. They may also be major selective factors in the evolution of species. Critical for the development of this body of theory so that it can be applied to real situations is quantitative information on the effects of these processes at the population level. Although it is probable that all three processes play a crucial role in the biology of the sticklebacks, even for this intensively studied group the quantitative data are lacking. Fortunately, the biology of the sticklebacks is such that they should provide such data from a wide geographical and habitat range, and from both experimental and observational studies.

9 POPULATION DYNAMICS

Introduction

Of the two fundamental characteristics of a population, one, the genetic composition, is discussed in Chapter 10. The other is numerical density. The abundance of a population of sticklebacks will depend on the rate at which new animals are added to the population by births or immigration and the rate at which fish are lost from the population by deaths and emigration. Sticklebacks are seasonal breeders, so the increase in density because of births in the population takes place only during a limited period of the year, whereas deaths take place throughout the year. In principle, immigration and emigration can also occur throughout the year, but the extent of such movements between populations is not known. Some studies on resident freshwater populations have suggested that movement in such populations can be very restricted (Hagen, 1967), but the migratory behaviour of anadromous populations makes possible some movement between them, particularly during the overwintering period in the coastal waters. Analysis of the genetic composition of anadromous populations from along the same coast may indicate the degree of mixing between the populations. Both birth and death rates and immigration and emigration rates also depend on the age structure of the population and its genetic composition. Thus, although the question, 'What determines the abundance of a population of sticklebacks?', looks deceptively simple, the answer involves many aspects of the biology of the stickleback.

The potential of a population to increase in abundance can be measured by its finite rate of increase, lambda (λ). For a population such as that of the stickleback, in which breeding is seasonal but there may be some overlap of the generations, λ is defined by the Euler-Lotka equation:

$$1 = \sum_{x=1}^{x=\infty} \lambda^{-x} \cdot l(x) \cdot m(x)$$

where $l(x)$ is the probability of an animal surviving to reach age x, and $m(x)$ is the average number of female zygotes produced by a female aged x. This expression assumes that $l(x)$ and $m(x)$ do not change with time and that neither survivorship, $l(x)$, nor fecundity, $m(x)$ depends on the density of the population. It also assumes that the population has a stable age distribution so that the proportion of the total population that is in each age class does not change as the absolute abundance changes. Measures of the mean fitness of the population and of the fitness of particular genotypes within the population can be derived from the Euler-Lotka equation, which provides one of the links between demographic and evolutionary studies of populations (Charlesworth, 1980).

The crucial components determining the value of λ are the age when the fish first reaches sexual maturity, the age-specific fecundity and the age-specific survivorship. The equation is given in terms of females only so the sex ratio of the population is also crucial in determining absolute abundance. Although there is some information on most of these components for stickleback populations, no single population has yet been studied in sufficient detail for estimates of λ to be made. Indeed, in spite of the attention that sticklebacks have received from biologists, there are few studies in which the abundance of a stickleback population has been measured over a period of years.

Age at Maturity and Lifespan

The method most commonly used to age fish, reading the number of annuli on the scales, cannot be used with sticklebacks because they lack scales. In some stickleback populations, the largest of the ear otoliths, the sagitta, can be used because it shows clear annual rings. An alternating sequence of opaque and transparent rings is laid down in the otolith, although the times of the year at which the two types of ring are laid down varies from population to population (Jones and Hynes, 1950; Allen and Wootton, 1982b). To age sticklebacks accurately using their otoliths, it is essential that samples be taken from the population over as much of the year as is practicable so that the otolith pattern for that population can be determined. In some populations, otoliths cannot be used either because the rings are not distinct or because, in addition to the annual rings, false rings are commonly present (Blouw and Hagen, 1981). Otoliths cannot be used without killing the fish, so the method must be supplemented by a less destructive method of ageing the fish.

Because sticklebacks are seasonal breeders, successive year classes are recruited into the population at approximately 1-year intervals, so that each age class has a 1-year advantage in growth over a younger age class. This situation leads to a multimodal length frequency distribution when the whole population is sampled. In theory the number of modes corresponds to the number of age classes. The use of length frequency distributions to determine the number of age classes in a stickleback population is complicated by two factors. The first is that the breeding season can be relatively long and several waves of young fish may be recruited into the population within 1 year (van Mullem, 1967), so the length frequency distribution for a single year class may be multimodal. A second problem is that once sexual maturity is reached, the growth rate tends to decline rapidly, with the result that the fastest growing fish of the youngest age class tend to catch up with the slowest growing adults and this leads to a blurring of the length frequency distribution. In many populations only two modes in the distribution can be recognised, the juveniles and the adults (Jones and Hynes, 1950). The adults may include more than one age class but this is not clearly shown in the distribution and must be confirmed by analysis of the otoliths. As with this latter analysis, frequent sampling of the population helps to clarify the ageing from length frequency data (van Mullem and van der Vlugt, 1964; Mann, 1971; Allen and Wootton, 1982b).

In the laboratory, with adequate care and the correct environmental conditions, *G. aculeatus* can be brought to sexual maturity within 6 months and may live for up to 5 years (Wootton, 1976), but in natural populations this potential is not realised. The environmental and physiological control of sexual maturation (Chapter 7) ensures that, in natural populations, the minimum age of maturity is 1 year, the fish breeding in the spring or summer that follows that of their birth. In many populations of *G. aculeatus*, this minimum age of maturity is achieved. Examples include the population in Llyn Frongoch, mid-Wales (Allen and Wootton, 1982b), the populations living in two streams in southern England (Mann, 1971), and in a stream in north-west England (Jones and Hynes, 1950). The anadromous populations found along the coast of The Netherlands and Belgium also reach sexual maturity at age 1 year (van Mullem and van der Vlugt, 1964). This tendency for populations of *G. aculeatus* to achieve the minimum age at maturity that is possible in the context of their seasonal mode of breeding is of interest, because reducing the age of first reproduction is a potent mechanism for increasing λ, the potential rate of increase of the population. Any further reduction in

the age of maturity would presumably result in the sticklebacks producing young at times of the year at which the offspring would have
little or no chance of surviving to become sexually active. Nevertheless, in some populations of *G. aculeatus* all or many of the fish do not
reach maturity until aged 2 years. This seems to be a feature of the
anadromous populations along the coast in parts of the Baltic Sea.
A study in the Asko Archipelago south of Stockholm found that
1-year-old fish reached a mean length of 31 mm but no sexually mature
fish less than 40 mm occurred. Sexual maturity was achieved at age 2
years, by which time the fish had reached a mean length of 55 mm
for females and 51 mm for males. The year-old fish do not move on
to the spawning ground in large numbers, indicating that they spend
the summer elsewhere along the coast (Aneer, 1973). A similar pattern
for northern anadromous populations, in which sexual maturity is
reached at 2 years and the 1-year-old fish spend summer in the sea
away from the spawning grounds, was described by Munzing (1959).
In the Alaskan lakes, Karluk and Bare, although a proportion of the
resident *G. aculeatus* populations did reach maturity at 1 year, a higher
proportion may not have matured until they were 2 years old
(Greenbank and Nelson, 1959). It is possible that a north-south cline
in the age of maturity is present, with northern populations maturing
at an older age than southern populations, but the data are too scanty
to confirm this.

Maximum lifespan of *G. aculeatus* shows considerable inter-
population variation. In many populations, few if any fish survive
to reach 2 years old after maturity has been reached at 1 year. This
pattern was shown by the Llyn Frongoch population (Allen and
Wootton, 1982b), the populations studied in southern England (Mann,
1971) and the anadromous population living along the coast of the Isle
of Tholen (The Netherlands) (van Mullem and van der Vlugt, 1964).
But in many other populations the maximum lifespan is greater. In Bare
and Karluk Lakes, Alaska, the lifespan was 2.25 years and a similar lifespan was reported for populations of the completely and partially
plated morphs living in two lakes in Quebec (Coad and Power, 1973b).
In Priddy Pool in south-west England, the population had a maximum
lifespan of nearly 4 years, similar to the lifespan for the population in
the River Birket in north-west England (Pennycuik, 1971a; Jones and
Hynes, 1950). A Baltic population included some fish that reached
4 years old, having become sexually mature at age 2 (Aneer, 1973).
The factors that account for this variation are not known, although
there may be a tendency for the maximum lifespan to be less in the

more southerly populations, but this would need to be confirmed by studies from many more populations.

Age at maturity and maximum lifespan for other species are similar to those of *G. aculeatus*, although *P. pungitius* females from a population in Lake Superior reached an age of 5 years and the males had a maximum span of 3 years (Griswold and Smith, 1973). At Isle Verte in the St. Lawrence Estuary, where four species are sympatric, analysis of the length frequency distributions suggested that *G. aculeatus* lives for up to 3 years, *G. wheatlandi* and *P. pungitius* for 1 to 2 years and *A. quadracus* for 2 to 3 years (Craig and FitzGerald, 1982). In all species the age of maturity was 1 year. Similar values have been reported for these four species at other sites in eastern Canada (Scott and Crossman, 1973; Wootton, 1976). *Spinachia spinachia* may be an annual fish, reaching maturity at 1 year old and dying after breeding (Jones and Hynes, 1950). Interestingly, *Aulorhynchus flavidus*, the tube-snout, which in external morphology so closely resembles *S. spinachia* (Chapter 3), may live for several years (Bayer, 1980).

Survivorship

Age-specific survivorship of sticklebacks has received virtually no systematic study. There are several anecdotal reports that mortality is particularly high after spawning, for example the corpses of adult *G. aculeatus* were described as littering the shallows of Lake Wapato in Washington State at the end of the breeding season (Hagen and Gilbertson, 1973a). The anadromous population of the Island of Tholen did show a sharp decline in the frequency of adults at the end of the breeding season such that no adults made the seaward migration (van Mullem and van der Vlugt, 1964). A similar decline in the frequency of adults in the population after spawning occurred in the Llyn Frongoch population (Allen and Wootton, 1982b). In neither of these studies was the survivorship rate estimated at different times of the year in order to indicate whether post-spawning survivorship was exceptionally low. Fish kept under good conditions in the laboratory will usually survive after spawning. The irreversible degenerative changes that follow reproduction in the truly semelparous Pacific salmon (*Oncorhynchus* spp.) have not yet been described in any stickleback population.

Age-specific Fecundity

Because of the relationship between body size and fecundity in sticklebacks (Chapter 7), the age-specific fecundity will depend on the size at which the females mature, and, if they survive to spawn in succeeding years, their growth during the time between periods of reproductive activity. There can be considerable inter-population variation in the size at maturity, even when the age at maturity is the same. This variation will partly reflect environmental differences, especially in food availability and temperature, but can also reflect genetic variation. When fertilised eggs from different populations of *G. aculeatus* from lakes on Vancouver Island were reared under identical conditions in the laboratory, there were significant differences between the populations in the minimum length at which the females matured. The lowest minimum standard length at maturity for a population was 35.2 mm and the highest was 48.4 mm. These differences observed in laboratory populations were strongly correlated with differences in the minimum length at maturity estimated from collections of fish taken from the lakes in the early part of the breeding season (McPhail, 1977). Although the fecundities for females from these populations were not measured, the differences in size at maturity probably also produced differences in fecundity.

A factor that complicates the estimation of age-specific fecundity of sticklebacks is the effect of food on the number of spawnings in a breeding season (Chapter 7). Thus females of the same length and age may produce very different numbers of eggs during a breeding season because of differences in their food supply.

Data from the population of *G. aculeatus* in Lake Wapato and experimental studies suggest that the fecundity of the females is sufficient for a male that successfully takes up a territory and builds a nest to produce several hundred free-swimming fry in a breeding season (Chapter 7) (Wootton, 1973a; Kynard, 1978a).

Sex Ratio

Outside the breeding season, the sex ratio of a stickleback population can only be determined accurately by dissecting the fish to reveal the gonads. During the breeding season, the territorial behaviour of the males and the schooling behaviour of females, juveniles and unsuccessful males make it difficult to estimate the sex ratio, because it is difficult to obtain a representative sample of the whole population. In the Baltic Sea south of Stockholm, collections of *G. aculeatus* made on the spawning grounds suggested that females outnumbered males. For the completely plated morph, the proportion of males to females was 0.206 in 1968 and 0.251 in 1969, and for the partially plated morph the proportions were 0.216 and 0.211 in 1968 and 1969, respectively (Aneer,

1973). Collections of the low plated morph from Llyn Frongoch also indicated a predominance of females. The proportion of males to females in a total sample of 209 fish taken over a year was 0.314 (Allen, 1980). Since the sex ratio of the zygotes at fertilisation would be expected to be 0.5, these observed sex ratios suggest that males suffer a higher mortality rate than females, but more data on sex ratios and changes in sex ratio with the age of each year class are required.

Population Dynamics in Lakes

Although there are few data on the population dynamics of stickle-backs, two reports have indicated their potential for numerical increase. Marion Lake is a small lake near Vancouver which has been the subject of an intensive ecological study. This indicated that although the lake contained two species of salmonid, the kokanee and the rainbow trout, it had no predators that utilised prey from the littoral zone where the water was less than 1 m deep. In 1974, 4000 *G. aculeatus* were introduced into the lake and in 1975 the population had reached 120 000, which provides an estimate of the finite rate of increase, λ, of 30 females per female per year (J.D. McPhail, pers. comm.). Lake Washington is an 87.6 km^2 lake close to Seattle, Washington State. Because of changes in sewage treatment, the lake has suffered episodes of eutrophication followed by some recovery. *Gasterosteus aculeatus* in the lake feed predominantly on zooplankton and form part of the limnetic fish community. In the period from 1968 to 1975, the estimated abundance of sticklebacks increased from very low numbers to about 3.3 million, and there has been a corresponding increase in the relative abundance of sticklebacks in the limnetic community which also includes juvenile sockeye salmon and the longfin smelt (Eggers *et al.*, 1978).

Lake Dalneye, which has an area of 1.36 km^2, has been studied for many years by Russian biologists who have estimated the total abundance of *G. aculeatus* at about 2.5 million, but with the sticklebacks divided between several populations each containing 100 000-200 000 fish (Krokhin, 1970). Population abundance has varied over a 30-year period, the variations being partly related to changes in the abundance of young sockeye salmon (Chapter 8). As in Lake Washington, the sticklebacks in Lake Dalneye are primarily consumers of zoo-plankton.

Population Dynamics in Rivers

In contrast to the large populations found in lakes, small populations occur in ponds and river backwaters. Since 1972, the abundance of a population of *G. aculeatus* living in a backwater of the River Rheidol near Aberystwyth has been estimated each October. The backwater has a surface area of about 190 m², but is slowly being filled in by heavy growths of vegetation. The estimations were made using the simple Petersen mark-recapture technique (Begon, 1979); the nature of the population is such that most of the assumptions of the method are probably not seriously violated. At each sampling, 400-500 fish were caught, so the population was probably estimated to within 25 per cent of its true size.

Abundance between 1972 and 1982 varied between 1450 and 5550 fish, and the estimated biomass varied from 670 g to 2350 g (Figure 9.1). The lowest abundance was in a year in which there was

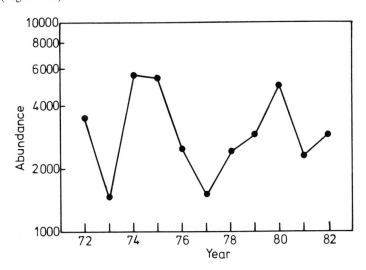

Figure 9.1: Petersen Estimates of Abundance of a Population of *G. aculeatus* in a Backwater of River Rheidol, Wales, 1972-82

unusually severe flooding in late summer just at the end of the breeding season. In October the population consists of two age classes, 0^+ and 1^+ fish, but the latter account for less than 15 per cent of the total population. Mortality between one October and the next for a year class was estimated as:

$$Z = \ln N(0) - \ln N(1)$$

where Z is the coefficient of mortality (p. 101), $N(0)$ is the number of 0^+ fish and $N(1)$ is the number of 1^+ fish. Z varied from 1.355 to 6.280, but was inversely correlated with $N(0)$, which suggests that the chances of a fish surviving from age 0^+ to 1^+ were less the greater the density of the 0^+ fish. In ecological terms, survivorship was density dependent. The mean length of the population in October was also inversely correlated with population density, as were the mean lengths of both the 0^+ and 1^+ fish in October. This indicates that growth was also density dependent.

The Rheidol population has survived two major environmental events. The flood in 1973 has already been mentioned, for in that year the population reached its lowest abundance, though in the following year the population reached its maximum abundance. The drought year of 1976 gave Britain its lowest rainfall for at least 200 years. Survivorship of fish between 1975 and 1976 was the poorest, so although numbers in the population in October 1976 were not unusually small, the proportion of 1^+ fish was.

Although this population has not been studied in sufficient detail to allow an adequate analysis of its population dynamics, the results do suggest that the potential rate of increase of a stickleback population is high enough for it to recover rapidly from a period of low numbers caused by unusual environmental conditions. But the results also suggest that further increase in abundance is limited by the population's own density. Experimental studies on the relationship between food supply and growth and reproduction suggest at least some of the mechanisms by which these density effects can operate (Figure 9.2). If, at high densities, there is less food available for each fish, then growth will be reduced and as a consequence so will the fecundity of the fish. Poor food supply during the breeding season will also reduce the number of spawnings by the females. In addition, the territorial behaviour of the males will restrict the number of breeding males that can be accommodated in the backwater. Experimental studies have confirmed that at high male densities a smaller proportion of the males successfully take up territories and build nests (Chapter 7). The small number of age classes present in stickleback populations may make them prone to catastrophic losses if a single age class suffers unusually high mortality. An example is provided by the collapse of the *G. aculeatus* population in Lake Wapato, probably because of high mortality suffered by the young-of-the-year (Chapter 8) (Kynard, 1979b).

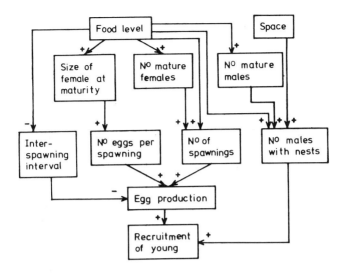

Figure 9.2: Block Diagram of Effect of Food Supply and Density of Males on Recruitment in *G. aculeatus*. After Stanley (1983)

Higher population densities than in the Rheidol were found in the River Weaver in north-west England when estimates were made in September and October. In ten samples, each of which represented a 25 m² area, the mean number of *G. aculeatus* was 1155, with a range from 601 to 1579 fish. The mean biomass was 400 g with a range from 203 to 614 g. *Pungitius pungitius* was collected in the same samples but had a lower abundance. The mean was 136 and the range from 19 to 399, and the mean biomass of *P. pungitius* was 51.7 g with a range from 6.3 to 164 g (Penczak and O'Hara, 1983).

Some anadromous populations of *G. aculeatus* can reach such high population densities that they support a fishery. Such fisheries have existed at various times in the Baltic Sea and along the Dutch coast, the fish being used for their oil, as manure or as feed for ducks (Wootton, 1976). A graphic picture of the abundance that sticklebacks can reach was given by Pennant in 1776:

Once in seven or eight years amazing shoals appear in the Welland and come up the river in the form of a vast column. The quantity is so great that they are used to manure the land and trials have been made to get oil from them.

There do not appear to be any other accounts that indicate that stickleback populations show regular cycles of abundance, although there was slight evidence of cyclic fluctuations in the *G. aculeatus* population in Priddy Pool (Pennycuik, 1971a). In this population, the cycle was interpreted as an effect of the parasite *Schistocephalus solidus* (Chapter 8).

Conclusion

Demographic studies of the stickleback are too few and too restricted for an adequate analysis of the population dynamics in different environments. The data do suggest that the potential for an increase in abundance is sufficiently high for a stickleback population to take advantage rapidly of any changes that are favourable to it. But the reproductive biology of the stickleback is such that there are likely to be effective restrictions placed on numerical increase as space or food comes to be in short supply. Predation, parasitism and inter-specific competition for food and space may also be effective factors restricting the abundance, but their relative importance in the demography of sticklebacks is unknown (Chapter 8). There is the possibility that at least some populations show regular cycles in abundance, a feature of animal populations that has generated considerable theoretical and empirical analysis.

10 ECOLOGICAL GENETICS

Introduction

If samples are taken from a population of *G. aculeatus* and carefully examined, the individual fish will differ in a number of easily measured characters such as the number of bony lateral plates. In some populations of *G. aculeatus*, three distinct patterns of arrangement of lateral plates are found, so fish belong to either the low, partially or completely plated morph (Figure 3.2) (Chapter 3). Even in populations in which only a single morph is found, individual fish vary in their number of plates. Other phenotypic characters, which vary both within and between populations, include the numbers of gill rakers, dorsal and anal fin rays and spines, and the extent of the development of spines and associated skeletal elements. One explanation for such variation within and between populations of the same species is that it is a result of the environmental conditions prevailing during the ontogeny of the individuals rather than any underlying genetic differences between individuals. If such phenotypic variations do reflect an underlying genetic variation, then a major problem is posed: what is the biological significance of the variation which is expressed phenotypically, but is generated by genetic variation in and between populations of the same species? Natural selection should result in the evolution of populations made up of individuals which, under the prevailing environmental circumstances, are most successful at leaving progeny. It might be expected that all individuals in a population would be uniform in their characteristics because those were the characteristics that resulted in reproductive success. Yet stickleback populations do not consist of numbers of virtually identical fish. On the contrary, *G. aculeatus* in particular is an extremely variable species. This situation is not unique to sticklebacks. Studies over the past two decades on protein variation in many species of plants and animals have confirmed that, far from being genetically uniform, natural populations contain high levels of genetic variation (Lewontin, 1974). Sticklebacks are a particularly favourable group for the study of the significance of this genetic variation, because they exhibit easily recognisable variation, are readily studied in both field and laboratory conditions and have a sufficiently short life history to make genetic analysis practicable.

Two major theories attempt to account for the genetic variation within populations and species. One theory argues that the bulk of the variation is present only because of historical accidents and does not have any biological significance. The variation shown by individuals does not affect their reproductive success significantly. The alternative theory argues that the variation in populations represents the effects of the action of natural selection, and that selection is the major process maintaining that variation. Both theories are used to account for the variation within and between stickleback populations.

Variation in *Gasterosteus aculeatus*

Lateral Plate Polymorphism

Introduction. Unfortunately, the biological variation represented by the plate morphs of *G. aculeatus* is reflected in the nomenclature used to discuss the phenomenon. European authors have often used the terms *trachurus*, *semiarmatus* and *leiurus* to describe the completely, partially and low plated morphs, respectively (Munzing, 1959, 1963; Wootton, 1976). North American authors have recently used *trachurus* to describe anadromous populations that overwinter in the sea and *leiurus* for populations resident in fresh water, irrespective of plate morphs. To avoid difficulties in this chapter, the terms *complete*, *partial* and *low* will be used to describe the three plate morphs, and *anadromous* and *resident* will be used to describe the lifestyle of the populations. Anadromous populations include those that breed in salt marshes or tidal pools.

Geographical Distribution of Plate Morphs. The worldwide distribution of the three morphs is so distinctive that it calls for some clear explanation but this has so far proved elusive (Figure 10.1).

In Europe, the anadromous populations that occur along the northern coastline, including the western and northern coastlines of Scandinavia and USSR as far as the White Sea and Novaya Zemla, the south-eastern coast of the Baltic and the coasts of Iceland and northern and eastern Scotland, contain only completes. Populations in the Black Sea also consist only of completes. There are differences in the degree of development of the lateral plates in these European populations of

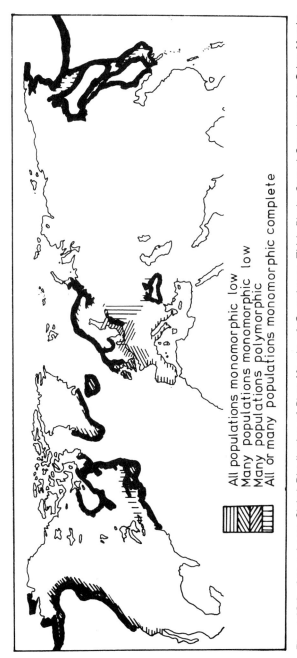

Figure 10.1: Approximate Global Distribution of Plate Morphs of *G. aculeatus*. Thick Black, Coastal Completes; for Other Morphs See Legend on Figure

All populations monomorphic low
Many populations monomorphic low
Many populations polymorphic
All or many populations monomorphic complete

the complete morph, for the plates are less strongly developed in the Baltic and Black Sea populations than they are in the northern coastal populations. Populations monomorphic for completes occur along the north-western and north-eastern coast of North America and the north-eastern coast of Asia. In much of the Baltic, the southern North Sea and the English Channel, the anadromous populations are polymorphic, containing completes, partials and lows. In these polymorphic populations the proportion of completes declines between the north and the south, so that in the western Baltic the completes form 70-90 per cent of the population, in the German Bight 50 per cent, off the Dutch and Belgium coasts 40 per cent, and in the western English Channel only 20 per cent. Further south, no strictly marine populations occur, although some coastal populations may live in relatively saline lagoons, but these populations contain only lows (Munzing, 1963; see also Wootton, 1976). Within the Baltic, the proportion of partials becomes high in the north-east, where they can form up to 80 per cent of a population (Gross, 1977). There is no comparable situation on either the east or west coast of North America. On the west coast, the anadromous populations, which reach as far south as Monterey Bay in California, consist almost entirely of completes, although a small percentage of incompletely plated fish may occur (Bell, 1976). A similar situation also holds for the east coast of North America (Hagen and Moodie, 1982) and probably for the east coast of Asia. Thus the only area in which partials form a significant proportion of the anadromous populations is in western Europe. The geographical area in which these polymorphic, anadromous populations occur is characterised by relatively shallow seas bordered by fertile, low-lying land crossed by innumerable slow-flowing rivers and streams. This is in marked contrast to the coastline of most of western America, where the coastal strip is narrow and carries fast-flowing rivers that rise in the bordering mountain ranges. In summary, anadromous populations are found in cold or relatively cool coastal waters and consist predominantly of completes.

The distribution of plate morphs in resident populations is much more complex. In much of western Europe, particularly in the south of this region, such freshwater populations contain only lows. East of the Oder river system, resident populations are monomorphic for completes. On the German Plain between the Elbe and the Oder-Neisse there is a complex situation. The anadromous populations contain all three morphs, but with the completes in the majority. Resident populations are either trimorphic, though the proportion of partials is usually low, or monomorphic for lows. Such monomorphic populations tend to occur in the peripheral areas of the drainage systems or in isolated lakes and ponds (Paepke, 1970, 1982; Munzing, 1972). On the west coast of North America, resident populations in southern California are monomorphic for lows, and in some of these populations the fish have no lateral plates. Further north, many resident populations contain only lows, but trimorphic and dimorphic populations also occur and a few populations are monomorphic for partials (Miller and Hubbs, 1969; Hagen and Gilbertson, 1972; Kynard and Curry, 1976). In contrast, on the east coast of America, the low morph is rare and most resident populations are either monomorphic containing only completes, or dimorphic with both completes and partials. An exception is that lows are not uncommon in populations on Newfoundland (Hagen and Moodie, 1982; Coad, 1983). Resident populations with high proportions of lows occur in Arctic Canada and in Greenland as well as in Iceland and Scandinavia, so this morph is not restricted only to relatively warm, southerly fresh waters. The distribution of morphs in Asia has not been thoroughly described recently, but resident populations containing lows occur in Kamchatka (Potapova, 1972) and also occur in southern Japan.

In summary, populations monomorphic for or containing a high proportion of completes are characteristic of cold coastal waters and the eastern regions of the northern continents. Populations with high proportions of lows tend to come from freshwaters in the southerly and westerly areas of the geographical range, but are also found well to the north.

Genetic Basis for the Plate Polymorphism. Breeding experiments using the three plate morphs have shown that the character is under genetic control. The breeding programmes have not yet been sufficiently intensive, nor have they used fish from a sufficient number of populations for a totally adequate model of the genetic system underlying the polymorphism to be proposed. Where crosses are made between parents that come from monomorphic populations of completes or lows, the results are usually clear cut. Crosses of complete X complete usually yield offspring that are all completes, and crosses of low X low produce offspring that are all lows. Complete X low crosses can produce an F_1 generation of partials, and the F_2 generation formed by crossing the F_1 partials consists of lows, partials and completes in a proportion not significantly different from 1:2:1. Such results suggest that the morphs are controlled by a single genetic locus with two alleles, T and t, possible at the locus. The genotype of completes would be *TT*, that of lows *tt*, and the partials would be the heterozygous genotype *Tt* (Munzing, 1959; Hagen and Gilbertson, 1973b). This model is not adequate because some crosses between completes and lows can give progeny that are all completes. Furthermore, populations that are monomorphic for partial are known and in such cases the cross of partial × partial can give progeny that are all partial, rather than the segregation for lows, partials and completes predicted by the model. Crosses between lows, partials and completes from the polymorphic Lake Wapato resident population gave further evidence that the single locus, two-allele model was inadequate (Hagen and Gilbertson, 1973b). Their results suggested a model in which the two genetic loci, 'A' and 'B', both with the two alleles, *A* or *a* and *B* or *b*, controlled plate morph. Morphs would be determined by the relative dosage (Table 10.1). The model predicts that populations monomorphic for partials would occur if there had been fixation of the alternative alleles at the two loci, so that only one genotype is present, either *AAbb* or *aaBB*.

Table 10.1: Model for the Relationship between Genotype and Phenotype for the Plate Morphs of *G. aculeatus* (Hagen and Gilbertson, 1973b)

		Genotype at locus A		
		AA	*Aa*	*aa*
	BB	Complete	Complete	Partial
Genotype at locus B	*Bb*	Complete	Partial	Low
	bb	Partial	Low	Low

At Friant in California, there is an unusual resident population which is essentially dimorphic for lows and completes. Crosses between lows and completes usually yielded only lows and completes in the F_1 generations, a result that is not predicted by the two-locus model but is compatible with a single-locus model in which the allele for complete is dominant over the allele for low. Crosses between fish from this population and from populations monomorphic for either lows or completes from two other sites in California did yield some partial progeny in the F_1 generations. These results suggested that the genetic basis for determining plate morph could vary between populations so that results obtained from crosses within populations would not necessarily be general for all populations of *G. aculeatus* (Avise, 1976).

Recent studies in the USSR on populations from the White Sea area suggest that the caudal keel with its associated plates, which is usually regarded as characteristic of the completes and partials, is determined by a separate genetic system from that of the plate morphs (Ziuganov, 1983). This suggestion raises the possibility that some populations that are apparently monomorphic for partials consist rather of a fourth, unusual morph, the 'keeled-low'.

Clearly, further analysis of the genetic control of plate morphs is required in which both intra- and inter-population crosses are made using material from as wide a geographic range as possible and the analysis is continued to the F_3 generation. But – and this cannot be overemphasised – equally clearly, plate morph is genetically controlled and is not determined by environmental factors operating during the ontogeny of individual sticklebacks. Such factors, especially temperature, do cause minor changes in characters such as plate number, but the effects are far too small to account for the differences between plate morphs (Lindsey, 1962).

The sequence of development of the lateral plates during the ontogeny of the complete morph is that the anterior plates start to appear first, then the plates associated with the caudal keel develop, and finally the gap between the anterior and caudal series is closed by the addition of plates to the posterior end of the anterior row and the front end of the caudal row. Row formation may not be complete until the fish is more than 30 mm long. This sequence is such that during ontogeny the complete morph passes through stages which resemble first the low and then the partial morphs. Within a geographical area, the relationship between the number of anterior and caudal plates in mature partials can be similar to the relationship in immature completes, which emphasises the similarity between an ontogenetic stage in the completes and the adult stage in the partials. An interpretation of these results is that partials and lows evolve from completes by paedomorphosis, that is, by the retention of pre-adult characters in adult phenotypes of descendant forms. The genes for plate morphs can then be interpreted as regulating the rate at which the lateral plates ossify. Completes carry alleles that specify rapid ossification, lows carry alleles for a low rate and partials arise from interactions between these alleles that produce a moderate rate of ossification (Bell, 1981). This model suggests that a relatively simple genetic change that results in a change in rate of formation of lateral plates can result in the evolution of lows or partials from an ancestral population of completes.

Fossil Records of Plate Morphs. A fossil of the completely plated morph was found in late Miocene/early Pliocene marine deposits in southern California (Bell, 1977). Fossils of the low morph are known from freshwater deposits from the Pliocene (Bell, 1973, 1974). This fossil record indicates that the plate polymorphism is at least 10 million years old and suggests that the ecological correlation between plate morph and salinity is equally old.

Evolutionary Relationship between Morphs. Most evidence suggests that the complete morph is primitive and the low and partial morphs have evolved from it. First, the geographical distribution and physiology of *Gasterosteus* indicates that it is basically a marine taxon that has invaded fresh water (Chapter 2). Secondly, in the Black Sea drainage, in which no low morphs have been recorded, two lakes contain populations of the partial morph. Lake Techirghiol, a saline lake in Romania, contains a population dimorphic for completes and partials. Lake Iznik, in north-western Turkey, has a population monomorphic for partials. The only plausible interpretation of these two populations is that they have evolved, by a process of reduction in the number of plates, from the complete morph after the lakes became isolated from the Black Sea. In the case of Lake Techirghiol, the final separation of the lake from the sea occurred only 100 years ago (Munzing, 1963). These populations also suggest that populations of partials and perhaps lows can evolve from populations of completes quite independently. The genotype of the complete morph seems to be such that the evolution of a reduction in plate number can occur readily and rapidly under certain conditions (Bell, 1976).

This suggests a model for evolution within *G. aculeatus*. The anadromous coastal populations of completes represent a stable genetic pool for the species. Variation within and between populations of anadromous completes seems to be much less at both a morphological and biochemical level than in resident populations (Hagen and Moodie, 1982). Resident populations which show considerable variation between and sometimes within populations have evolved from this core genetic unit (Bell, 1976). A complication to this model is that even fish from resident populations show considerable tolerance to changes in salinity (Chapters 2 and 5), and in some places breed in saline coastal lagoons (Bell, 1979). It is debatable to what extent the resident populations have been derived independently from an ancestral anadromous population of completes or by watershed-to-watershed colonisation by strays making relatively short sea migrations from resident populations.

The most parsimonious hypothesis, but probably the least likely, is that the evolution of a resident population from an anadromous population occurred only once and that the subsequent distribution of resident populations has been by strays from the original resident population. In other words, the evolution of the low morph from the complete morph took place only once. A detailed analysis of the genetic similarities between resident populations over the whole geographical range of *G. aculeatus* is urgently required to indicate the extent to which these populations have been derived by independent but parallel evolution from an ancestral stock of the complete morph or by watershed colonisation from a small number of ancestral resident populations. Many anadromous populations in western Europe do contain a small proportion of the lows and these could be potential colonisers of freshwater systems (Munzing, 1963).

Partials present a particular problem because the evidence suggests that two mechanisms can produce significant proportions of this morph in populations. Crosses between completes and lows usually give offspring that are partials, so in some populations the partials represent hybrids or intergrades produced by matings between completes and lows. In other populations, the partials may be monomorphic or present in frequencies too high to be accounted for by hybridisation. In these populations the partial represents a self-sustaining genotype in its own right (Hagen and Gilbertson, 1973b; Hagen and Moodie, 1982).

Significance of the Distribution of Plate Morphs. To what extent does the present-day distribution of plate morphs (Figure 10.1) represent the result of historical accidents, particularly those that stem from the Pleistocene glaciations, and to what extent is it the consequence of natural selection operating on the morphs or other characters that are correlated with the plate character? The global distribution suggests that the abiotic factors of temperature and salinity are important selective factors (Hagen and Moodie, 1982). Lows are largely excluded from cold sea water and also from fresh waters in areas that have a continental climate with a wide temperature range between winter and summer. They do occur in fresh waters in areas with cold winters such as Greenland, but only where those fresh waters are close to the sea. These differences in the global distribution correlate with differences in the physiological tolerances and mechanisms of osmoregulation found in a few populations of completes and lows (Chapter 5). The experimental analysis of the interactions between the temperature, osmotic concentration and ionic composition of the water in determining the survivorship and reproductive success of sticklebacks attracted much attention four decades ago, but interest waned as more attention was paid to behavioural biology. As a result, there is an inadequate physiological background against which hypotheses on the causation of the global geographical pattern can be assessed. It is not clear whether the correlation between plate morph and salinity tolerance is because of a close linkage between the genes responsible for these characters or because the plates themselves play some role in osmoregulation by acting as a store of calcium from which these ions can be reabsorbed into the blood if serum levels of calcium drop too low (Chapter 2).

Nevertheless, the global distribution of the three morphs is such that it is unlikely that the absence of a particular morph from an area is a result of the historical accident that the morph has failed to disperse into the area. Rather, the absence of a morph indicates that it lacks the necessary adaptations to establish itself successfully in the area.

Within geographical areas, the relative role of historical accident and natural selection in determining the distribution of morphs is less obvious.

The most striking feature of the distribution of morphs in western Europe is the presence of significant proportions of partials in populations from the North German Plain and the lowlands that border the southern North Sea. This pattern is usually interpreted as being a result of extensive intergradation between completes and lows in this area, which at various times during the Pleistocene glaciations formed geographically separate populations. It is assumed that at the height of the advances of the ice sheet, populations of lows were forced into the relatively warm areas in south-western Europe, whereas populations of completes were found in the cold seas off the north-west coast and also probably in the south-east around the Black Sea. As the ice retreated, these separated populations tended to converge in the area of the North Sea and northern Germany where extensive hybridisation took place resulting in both anadromous and resident trimorphic populations containing significant proportions of partials. The limits on the area in which trimorphic populations occurred were set by the physiological limitations of the lows and completes with respect to temperature and salinity. This interpretation essentially assumes that within the area of hybridisation the hybrid morph (partials) are not at any selective disadvantage in comparison with the parental morphs (lows and completes). The proportion of the three morphs in a population is a reflection of the proportion of parental morphs, which is determined by their pattern of dispersion into the area of sympatry and their physiological tolerances (Munzing, 1959, 1963; Paepke, 1970).

Recent studies on populations from the North German Plain indicate that, at least in some resident populations, the partials are at a disadvantage. Crosses between lows and completes from trimorphic populations yielded progeny that consisted of lows, partials and completes, but the partials were relatively under-represented. Partials that were viable grew as fast as lows or completes but tended to reach sexual maturity later and to have a shorter lifespan. In these populations, the partials typically form only a small proportion of the total (Paepke, 1982). In contrast, in anadromous populations of the North Sea, the proportions of partials are high. For example, in populations from the Elbe estuary they account for about 40 per cent of the population (completes about 55 per cent, lows about 5 per cent) (Munzing, 1963), so in this area they do not seem to be at any selective disadvantage. Indeed, such high proportions of a hybrid (heterozygous) morph suggest that the partials are at some selective advantage. Unfortunately, the genetic analysis of these trimorphic populations has received relatively little attention, so it is possible that in at least some of these populations the partials are not simply hybrids but represent a stable genotype in their own right. The relative proportions of the partials in a population would then be a reflection of some, as yet unknown, selective factors. In these anadromous populations, the juveniles within a river system assemble in large schools during late summer and autumn and migrate to the sea, where they are likely to mix with juveniles from adjacent river systems so the potential for gene flow between populations is considerable (Munzing, 1963).

Resident populations in eastern Canada contrast strongly with those in western Europe because they are mostly monomorphic for the complete morph or are dimorphic for completes and partials, with the completes typically forming over 90 per cent of the population. A few populations which are monomorphic for partials also occur, and the few populations found that are dimorphic with high proportions of partials are in lakes and their tributaries close to the sea (Hagen and Moodie, 1982; Coad, 1983). Since the low morph is largely absent from the area, the partials cannot be maintained by hybridisation, but have presumably evolved from completes by reduction in the body armature independently of other populations of partials. There is a weak correlation between climate and the distribution of resident populations monomorphic for or containing a high proportion of partials which are not maintained by hybridisation. Such populations tend to occur in areas such as eastern Canada and Alaska where there is a relatively wide fluctuation in the annual temperature, with cold winters (Hagen and Moodie, 1982).

On the Pacific coast of North America, the resident populations can be monomorphic for any of the three morphs, trimorphic, or dimorphic (Hagen and Gilbertson, 1972; Avise, 1976; Kynard and Curry, 1976). One interpretation of this diversity is that it reflects the results of hybridisation between anadromous completes and resident lows, together with some introgression of genes between populations of completes and lows. For example, the introgression of genes from a population of completes into a population of lows would tend to increase the plate number characteristic of that population of lows (Miller and Hubbs, 1969). This interpretation suggests that inter-population differences largely reflect historical accidents which have allowed intergradation and introgression between some populations. However, there is now considerable circumstantial evidence that selection has played a major role (Hagen and McPhail, 1970; Bell, 1976), although the relevant selective factors have proved elusive.

A survey of resident populations from the Pacific coast between Alaska and Washington State found that there were wide variations in the frequencies of the plate morphs between populations. There was no correlation between the frequency of plate morphs and such obvious environmental variables as temperature, pH, distance to the sea or total dissolved solids (TDS), with the slight exception that the proportion of partials in a population tended to decrease as TDS increased. Even this weak correlation is difficult to interpret because the range of TDS recorded was limited. There was some tendency for large oligotrophic lakes to contain polymorphic populations (Hagen and Gilbertson, 1972). Interestingly, the survey suggested that resident populations from lakes that did not come into contact with anadromous populations resembled the latter more than did stream populations that were sympatric with the anadromous populations during the breeding season. Selective factors in a large lake may, in some respects, more closely resemble those in coastal waters than those in streams.

Stronger evidence for the importance of selection in maintaining the plate polymorphism came from surveys within limited geographical areas. In the Chehalis River in Washington State, the frequency of plate morphs changed dramatically within a few kilometres. Some resident populations were monomorphic for completes, some for lows and others polymorphic. The populations that were monomorphic for completes were many kilometres from the sea and the anadromous completes of this coastline (Hagen and Gilbertson, 1972). Brush Creek is a small drainage system in northern California in which the three plate morphs occur. In the fast-flowing clear waters that characterised the upper drainage, the population was monomorphic for completes. In the lower drainage, where the stream was sluggish and the water deeply stained, the population contained a high proportion of lows, although all three morphs were present. A shallow lagoon near the mouth of the drainage contained the three morphs in relatively similar proportions, whereas in an isolated pool close to Brush Creek the population was virtually monomorphic for lows. The correlation within this drainage system between habitat and morph frequency strongly suggests that selection is maintaining the frequencies. In other areas of California, a correlation between morph frequency and stream gradient similar to that in Brush Creek has been observed (Bell, 1982a).

Further evidence for importance of selection in maintaining morph frequencies emerged from a study of a trimorphic population that was introduced, in 1967, into Heisholt Lake on Texada Island off the coast of British Columbia (Maclean, 1980). Sampling of this population between 1970 and 1973 suggested that the frequencies of completes, partials and lows varied between areas of the lake, but that the frequencies were not related to the increase in population density that occurred during the period. MacLean argued that the population was divided into 'residents', fish that held territories or home ranges both during the breeding season and after it, and 'non-residents', which moved from area to area in schools. Changes in morph frequency would reflect the relative success of the morphs in competition for territories.

Reproductive Isolation between Morphs. If, as the evidence presented above suggests, the plate morph does have an adaptive significance such that the three morphs are adapted to different environments, the genetic integrity of each morph is in danger of being broken down by gene flow between them. Evidence that there can be strong barriers to gene flow between morphs that are sympatric came from a study of the Little Campbell River in south-western British Columbia (Hagen, 1967). In this river there is a resident population monomorphic for lows, and in summer an anadromous population breeds in the lower reaches. Samples taken along the stream from its mouth to the headwaters showed that there was only a restricted zone, between 1 and 2 km in length, in which evidence of hybridisation between the anadromous completes and the resident lows could be found. Within the hybrid zone, completes, lows and partials occurred. Immediately upstream of the zone, the population was monomorphic for lows and in this area the lows had the smallest mean number of lateral plates shown along the length of the stream (Figure 10.2). This suggests that there was

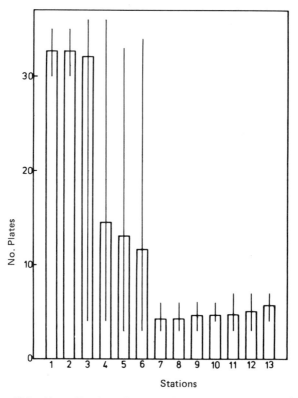

Figure 10.2: Mean Number of Lateral Plates in Population of *G. aculeatus* during Summer in Little Campbell River, British Columbia. Stations Ordered in Increasing Distance from the River's Mouth. Range at Each Station Given by thin Vertical Line. Data from Hagen (1967)

no introgression of genes from the population of completes into the population of lows. On the contrary, it suggests that there was strong selection against such introgression. Downstream of the hybrid zone, the population was monomorphic for completes. Several factors contributed to the reproductive isolation between the anadromous and resident populations. The anadromous population did not start breeding until late in the breeding season of the lows. The residents tended to nest in still, discoloured water on a muddy substrate. The anadromous population preferred to nest on a sandy substrate. Experiments in which individuals from the anadromous population were transferred to the habitat typical of the residents suggested that the former were poorly adapted to the residents' habitat. In the laboratory, the two morphs readily interbred and there was no evidence that under laboratory conditions the hybrids were at a disadvantage, but in the natural habitat it is likely that the hybrids were at a selective disadvantage in the typical habitats of both the resident and anadromous populations. An analysis of the courtship showed that both males and females tended to choose mates from their own population, that is they showed positive assortative mating. The choices were not absolute: in just over 60 per cent of the tests, a mate belonging to the same population was chosen in preference to one from the other population (Hay and McPhail, 1975). During courtship, there were differences between the resident and anadromous populations. Males from the resident population tended to zig-zag and glue more than males from the anadromous population (Chapter 7). An anadromous male was more likely to bite the female on his first approach to her. Resident males courted resident females more vigorously than anadromous females, but the anadromous males showed no difference in their courtship (McPhail and Hay, 1983). No other population has been studied in such detail as that in the Little Campbell, but studies on the distribution of plate morphs in Belgium also suggested the presence of a narrow hybrid zone between the resident and anadromous populations, which indicates that selective factors favour reproductive isolation between the populations (Heuts, 1947).

Conclusion. A final decision on the relative importance of selection and historical accident in determining the present distribution of plate morphs is not possible. Whenever detailed studies within a drainage system are carried out, strong evidence that the distribution of plate morphs is a result of the action of selective factors has emerged. However, in many of these systems it is possible that the presence of the plate morphs in the system has depended on historical accidents, particularly those associated with the changes in temperature and sea level that characterised the complex succession of glacial and interglacial periods of the Pleistocene. Once the morphs had arrived in the system, natural selection would then adjust their relative abundances and distributions. An understanding of the factors that maintain polymorphisms in natural populations is a major problem for population biologists and the present failure to account convincingly for the plate polymorphism in *G. aculeatus* is an accurate reflection of that problem.

Variation in Lateral Plate Number

Genetic Basis of Variation in Plate Number. Even within a plate morph, there is both inter- and intra-population variation in the number of lateral plates. Some of this variation is a result of differences in temperature during the ontogeny of individuals (Lindsey, 1962), but a significant proportion reflects genetic variation. The number of plates is under polygenic control so the relative importance of genetic variation in determining variation in a particular population and for a specified set of environmental conditions can be estimated by the heritability of plate number. Heritability (in the narrow sense) is defined as the ratio of additive genetic variance to the total phenotypic variance (Falconer, 1981). If, within a population, all the variation in plate number was a result of external environmental factors and there was no underlying genetic variation, the heritability would be 0.0, whereas if all the variation was due to genetic variation in the population and none to the effects of environmental factors on the phenotype, heritability would be 1.0. A high value of heritability for a character indicates that there is available a large store of additive genetic variance in the population, allowing the character to respond rapidly to selection.

Heritabilities for plate number in the Lake Wapato population were estimated as 0.50 at 21°C and 0.83 at 16°C. Some fish have a different number of plates on each flank and this asymmetry had a heritability of 0.63. These high values for heritability suggest that plate number and plate asymmetry could respond rapidly to selection, but they have not been subject to directional selection in the past. Although these estimates were obtained from a single population, it is possible that the results can be generalised to other populations (Hagen, 1973).

Geographical Variation in Plate Number. Although plate number does vary in anadromous populations of completes, this variation has not yet been studied systematically. The much more obvious variation in plate numbers in partials and lows has attracted attention. Again, the major problem is to determine to what extent the variation reflects the effect of selective factors and to what extent that of historical accidents which have allowed populations to become isolated or to regain contact with other stickleback populations.

On the basis of a survey of populations from the Pacific coast of America, Miller and Hubbs (1969) argued that the variation in plate number could largely be explained by intergradation and introgression between populations, with, in some cases, periods of isolation interrupted by periods in which gene flow between populations was resumed. Introgression from anadromous populations of completes into populations of lows resulted in the latter having a relatively high number of lateral plates. Intergradation and introgression between typical populations of resident lows and the populations in southern California which lack or have a small number of lateral plates resulted in a decrease in the mean number of plates in the lows.

This central role of intergradation and introgression in determining variation in plate number has been called into question by studies both of European and American populations.

A survey that covered all of Europe of the geographical variation in plate number in resident populations of lows showed that there was significant clinal variation. The cline followed an inverted U-shape from north to south. Populations from the centre of the distribution had the highest mean plate count whereas the lowest plate counts tended to be in peripheral populations in the north-west and the south. In these northern and southern peripheral populations, some were characterised by a high proportion of fish with no plates. There was also a tendency for plate number to increase from west to east. No abiotic factors could easily account for the shape of the cline, but the pattern could be related to the geographical distribution of two major fish predators of resident sticklebacks, pike and perch (Chapter 8). Highest plate numbers coincided with areas where the density of these predators was probably highest (Gross, 1977).

The detailed study of the populations in the Little Campbell River in British Columbia suggested that intergradation and introgression were unlikely to be important in determining variation in plate number because of the effective reproductive isolation between populations of completes and lows (Hagen, 1967).

A survey of resident populations from the Pacific coast between Alaska and Washington State found that populations of lows and partials showed two obvious frequency distributions for the number of plates. In some populations there was a strong mode for seven plates along each flank, whereas in other populations there was no distinct modal class and few or no fish with seven or more plates. On the Queen Charlotte Islands, ten lakes contained populations that had a modal number of seven plates. The common environmental factor linking those populations that show a mode at seven plates is the presence of predatory fish (Hagen and Gilbertson, 1972; Moodie and Reimchen, 1976). This correlation suggests that fish predation is a selective factor favouring sticklebacks with seven plates.

The population in Lake Wapato was trimorphic but with a very high proportion of lows which showed a strong mode at seven plates. A comparison of the plate counts for fish that were taken from the stomachs of trout compared with those caught in a seine net suggested that the trout preyed selectively on certain plate phenotypes. Fish with less than seven or more than eight plates were more likely to be predated than fish with seven plates. Over one year, 1968-69, the proportion of fish with seven plates in the population increased from 56 per cent to 65 per cent, suggesting strong selection. Because the population was annual, a minimum estimate of selection within a year class could be obtained by comparing the frequency of plate phenotypes present in the autumn with the frequency present in the same cohort of fish in the following June. This also showed selection for particular plate morphs, but in this case the optimum phenotype had a total plate count of 17 (8 + 9) (Hagen and Gilbertson, 1973a). When the population declined abruptly as the population of pumpkin-seed sunfish increased (Chapter 8), the decline was accompanied by a change in the frequency of plate phenotypes. In terms of total plate counts, phenotypes with 13 or fewer plates declined, 14 remained constant, 15 to 18 increased but 19 decreased. These changes suggest that the decline was associated with stabilising selection for fish with an intermediate number of plates (Kynard, 1979b).

Mayer Lake, on the Queen Charlotte Islands, contains a population of lows that reach an unusually large adult size and are melanic. This population also has a mode at seven plates. When the plate counts of fish taken from the stomachs of cut-throat trout were compared with those of fish caught by seining, there was a significant excess of fish with eight plates in the fish stomachs (Moodie, 1972b).

Ventura River is a drainage system in southern California which contains populations of lows. The system is divided into an upper and lower zone by an area that is usually dry but is subject to unpredictable flooding. The upper reaches contain rainbow trout. In the lower reaches the commonest plate phenotype had a total of ten plates (usually 5 + 5), whereas in the upper reaches the commonest phenotype was 14 (usually 7 + 7); a sharp cline coinciding with the dry area separated the two populations (Figure 10.3). There was some evidence of slight

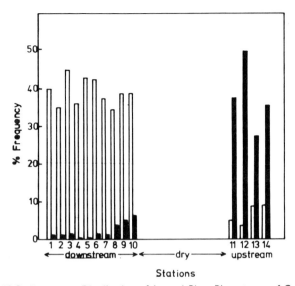

Figure 10.3: Frequency Distribution of Lateral Plate Phenotypes of *G. aculeatus* in Downstream and Upstream Portions of Ventura River Drainage (California). Open Bars, 10 Plates; Black Bars, 14 Plates. Stations Ordered in Increasing Distance from River Mouth. Rainbow Trout Present Only in Upstream Portion. Data from Bell and Richkind (1981)

gene flow from the upstream population into the downstream one, a flow correlated with the direction of water flow. One interpretation of this pattern is that the selection pressure imposed by the predatory trout has led to the development of the cline in the population in the drainage system. This cline developed in the face of gene flow within the drainage system and without an extensive period of isolation between the populations in the upper and lower reaches, that is, the cline developed parapatrically (Endler, 1977; Bell and Richkind, 1981).

These studies suggest that fish predators act as selective factors on plate number because they prey selectively on certain plate phenotypes. Field evidence from Lakes Wapato and Mayer support this hypothesis. Laboratory experiments have also supported the hypothesis but have indicated that the interaction between predator and plate phenotype can be complex. Squawfish and cut-throat trout were used as predators in a series of independent studies (Moodie *et al.*, 1973). Tests carried out under 'winter' conditions (low temperatures and short photoperiod) showed that the plate phenotype 7 + 7 was less liable to be eaten, but under 'summer' conditions significantly more 7 + 7 fish were eaten. Under 'summer' conditions, the 7 + 7 fish tended to leave the cover of vegetation sooner than the 4 + 4 or 5 + 5 phenotypes present in the same experiment.

In southern California, predatory fish are relatively rare but garter snakes do predate sticklebacks (Chapter 8). Experiments showed that the plate phenotype 5 + 5 had the highest rate of survival during snake predation (Bell and Haglund, 1978), so other predators may also act as selective factors on plate phenotype.

Reimchen (1983) showed that the anterior lateral plates which abut the basal plates supporting the first and second dorsal spines, together with the lateral plate which lies above the ascending process of the pelvic skeleton, are particularly important in providing structural support for the erect dorsal and pelvic spines. This may provide a mechanical explanation for the relative advantage of some plate phenotypes under predation. There is also some evidence that characters correlated with plate phenotype are involved. Tests in which larvae were exposed to predation before the plates had developed produced a higher frequency of seven-plated fish (Moodie *et al.*, 1973), It is possible that behavioural traits closely correlated with plate phenotype are important. In European populations of resident lows, there is some evidence that fish with a relatively high number of plates are at an advantage under predation (Gross, 1977).

Other behavioural traits also seem to be correlated with plate phenotype. Studies on the reproductive biology of males in Lake Wapato suggested that males that nested in deeper water and in cover were more successful than other males. In experiments, males with a total count of 16 plates were outcompeted for concealed nest sites by males with 14 plates. They were also less successful in obtaining nest sites or rearing young than males with 14 or 15 plates. Males with 16 plates tended to nest in the shallowest water in Lake Wapato. This plate phenotype may have had some advantage outside the breeding season (Kynard, 1979a). During the breeding season in Mayer Lake, plate phenotypes differed in the depth of water in which they nested and in the distance from cover that the nests were built. The 7 + 7 and 6 + 7 phenotypes were in deeper water closer to vegetation, two qualities of a nest site that seem to improve reproductive success (Chapter 7) (Moodie, 1972a,b). Plate phenotypes may also differ in their aggressiveness (Huntingford, 1981).

All these studies suggest that variation in lateral plate number is not a character that is selectively neutral but, either directly or because there is a close correlation between plate phenotype and behavioural characteristics, is sensitive to selective factors. Inter-population variation in plate number is in many cases probably due to differences in selection and not to differences in the amount of intergradation and introgression. However, the factors that maintain variation in plate number within a population are still completely unknown. A much more detailed study of the genetic control of the variation and the relationship between plate phenotype and the growth, reproduction and survival characteristics of the phenotypes is required.

Plateless Populations. Resident populations in which all or a high proportion of the individuals lack plates are now known from several geographical areas. These populations are typically on the southern or western periphery of the continental distributions of *G. aculeatus*. On the Pacific coast of America, they occur on the Queen Charlotte Islands, on Texada Island, which lies between Vancouver Island and the mainland, and in southern California (Bell, 1976). The Californian populations are frequently given sub-specific status as *G. aculeatus williamsoni*. In Europe, plateless fish are found in populations in the Outer Hebrides (Scotland) and in Italy (Campbell, 1979; Giles, 1983a; Bianco, 1980). Plateless *Gasterosteus* are known as fossils from the late Miocene deposits in Nevada, 10 million years old (Bell, 1974). This wide geographical distribution of isolated populations and the fossil evidence strongly suggest that the independent and parallel evolution of platelessness has taken place on several occasions (Bell, 1974). The selective factors that lead to this condition are not known. It is frequently associated with a general reduction in the development of the skeleton so that plateless fish often have relatively small spines and in some populations there may be a reduction and loss of dorsal and pelvic spines.

Variation in Defensive Apparatus

Spine Length. Two geographical surveys of variation in spine length showed that length was correlated with the presence of predators. The lengths of the dorsal and pelvic spines were also correlated. On the basis of the distribution of potential predators of the stickleback in Europe, Gross (1978) assumed that the geographical variation in predator pressure was small in the marine environment, but that in fresh water, predator pressure declined from a maximum in central Europe to a minimum both in northern and southern regions. In marine populations of sticklebacks, mean dorsal spine length changed little towards the north, increased slightly towards the east but decreased strongly in the south-east. In resident populations of lows, there was a symmetrical arched cline in spine length from north to south, so that populations in the centre of the European distribution had the longest spines, and those on the northern and southern peripheries the shortest. When comparisons were made for fish of similar lengths, the mean spine length in marine populations was longer than in resident populations subject to predation, which in turn had significantly longer spines than resident populations not subject to predation. For a fish of length 50 mm, the mean length of the second dorsal spine was: marine, 4.9 mm; resident predated, 4.3 mm; and resident unpredated, 3.9 mm.

Along the Pacific coast of America, the mean length of the pelvic spines was significantly greater in resident populations where fish predators were common. There was also a significant correlation between mean spine length and the frequency of seven-plated fish (Hagen and Gilbertson, 1972). Within the restricted area of the Queen Charlotte Islands, the relative length of the pelvic spines tended to be greater in populations subject to predation although the difference was not statistically significant (Moodie and Reimchen, 1976). Females from Mayer Lake that were taken from the stomachs of cut-throat trout had significantly shorter pelvic spines than fish collected by seine nets in the same area (Moodie, 1972b). The Mayer Lake sticklebacks had relatively longer pelvic spines than populations of typical, resident lows that lived in the heavily weeded, muddy, stained streams that feed and drain the lake. The streams did not contain predators. Interestingly, the stream populations did not have the mode for seven-plates shown by the Mayer Lake population. These phenotypic differences were between populations separated by only a few metres (Moodie, 1972a,b).

Pelvic Skeleton and Spine Reduction. In Europe, the length of the pelvic spine and the relative development of the pelvic skeleton shows a similar pattern of geographical variation to the length of the dorsal spines (Gross, 1978). Both in Europe and in North America, there are a few populations that show a reduction in the development of the pelvic skeleton and spines. The presence of dorsal and pelvic spines is such a characteristic feature of *G. aculeatus* that the discovery of populations in which such spines are reduced or absent emphasises how variable this taxon is. As with the absence of lateral plates, the reduction of the pelvic skeleton and loss of spines seem to occur in western and southern peripheral populations and the two types of skeletal reduction are often associated (Figure 10.4), as they were in the plateless fossil sticklebacks described earlier.

On the Queen Charlotte Islands, there are several populations show-ing reduction in the pelvic skeleton and in the number of spines. Typically, they occur in small, shallow lakes with a low pH and situated in areas of sphagnum bog. Similar populations occur in equivalent habitats in the Outer Hebrides off the west coast of Scotland (Reimchen, 1980; Campbell, 1979; Giles, 1983a). On Texada Island, well to the south of the Queen Charlottes, Paxton Lake contains a population in which the pelvic skeleton and spines are sometimes absent, as are the lateral plates (Bell, 1974; Larson, 1976).

The selective factors that have led to the evolution of the spine-deficient phenotype are not known. One hypothesis is that the reduc-tion takes place where predators are absent or at a low density (Bell, 1974). This reduction would allow a redistribution of the resources invested in defensive apparatus into other areas such as reproduction (Gross, 1977). This hypothesis was made more unlikely by the observa-tion that the spine-deficient populations on the Outer Hebrides were subject to both bird and fish predation (Giles, 1983a).

A population in Boulton Lake on the Queen Charlottes is predated by birds but not fish. About 80 per cent of the population lack the second dorsal spine, 65 per cent have only a vestige of the pelvic spine, and 20 per cent have a complete pelvic skeleton including the spines. The phenotype with the greatest spine deficiency is the most common, and only 4.3 per cent of the population have a full complement of dorsal and pelvic spines. These frequencies have remained stable for at least a decade. The phenotypes and sexes in Boulton Lake show some differences in their habitats. Adult females were found more frequently in open water than males, which predominated in benthic and shoreline areas. Females also tended to be more spined than males. Among the sub-adults of both males and females, those without pelvic spines tended to be more common near the shore. Small juveniles, in which the spines have not yet developed, were also found primarily near the shore. It is possible that the fish in the more open waters are more susceptible to bird predation and some development of the spines is an advantage, whereas those inshore are more prone to predation by invertebrates, a situation in which spines may be a disadvantage by pro-viding protrusions that the predator can grasp (Chapter 8) (Reimchen, 1980; Reist, 1980b).

Paxton Lake contains two major phenotypes. One is a relatively

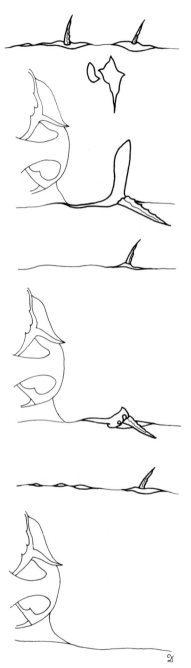

Figure 10.4: Examples of Spine and Pelvic Skeleton Reduction in *G. aculeatus* from Outer Hebrides (cf. Figure 8.3). Drawn from Alizirin-red Stained Specimens Supplied by N. Giles

typical phenotype for the low morph, with a full complement of spines and nine to fourteen lateral plates. The other phenotype shows variable development of the pelvic skeleton, spines and lateral plates (Bell, 1974). The two phenotypes differ in several respects, which suggests that the spine-deficient morph is adapted to a benthic mode of life whereas the typical morph is better adapted to a life in open water (Larson, 1976). The origin of these two coexisting populations is not known, nor is there information on the amount of gene flow between them within this small (17 ha) lake.

Impressed by the correlation between the incidence of spine-deficient populations and water of low pH and low calcium content on the Outer Hebrides, Giles (1983a) suggested that the evolution of the spine-deficient condition was related to calcium deficiency. Calcium plays an important role in osmoregulation of sticklebacks in fresh water (Chapter 2), and a well developed skeleton will tie up high concentrations of calcium in conditions in which the fish is constantly losing calcium to the external environment. Skeletal reduction may reduce the amount of calcium the fish has to retrieve from the surrounding water. Clearly, the interrelationships between osmoregulation and calcium in spine-deficient and normal morphs is open to experimental analysis. Loss of calcium when the eggs are laid may also be a factor that favours economising on calcium in waters in which calcium is difficult to obtain.

A reduction in the spines and in the anterior lateral plates may improve the streamlining and manoeuvrability of the fish, increasing the behavioural capacity to escape from predators at the cost of a reduction in the morphological defences (F.A. Huntingford, pers. comm.). Variable skeletal reduction may represent a compromise, given a complex combination of selection for economy in the use of calcium and selection for some morphological and behavioural defences against bird and fish predation.

Variation in Gill Rakers

In the Little Campbell River the anadromous completes had a significantly higher mean number of rakers than the resident lows. Crosses within and between morphs suggested that this difference was under genetic control (Hagen, 1967). Heritability for raker number in the Lake Wapato population was estimated at 0.63 at 21°C (Hagen, 1973). A survey of resident populations along the Pacific coast showed that the mean number of rakers on the first arch tended to cluster at high values, 19 to 21.7, or at low values, 14.6 to 17. Mean number could vary over a short distance. In Mayer Lake, the mean was 21.2 whereas in its streams it was only 16.6. The spine-deficient benthic morph of Paxton Lake had a significantly lower average number of rakers than the limnetic form (Larson, 1976). High raker number is associated with feeding in relatively open waters in which zooplankton forms a significant part of the diet, whereas a low number is associated with benthic feeding on larger prey items (Chapter 4). In general, the mean raker number of populations resident in lakes was more like that of marine anadromous populations than of resident populations of streams, even when the latter shared the same river system as an anadromous population, allowing the possibility of gene flow. This suggests that the selective factors that influence raker number are more similar in lakes and in the sea than they are in streams. Convergence in raker number between lake and marine populations is the result of selection, not intergradation and introgression (Hagen and Gilbertson, 1972).

Variation in Body Shape and Size

There is evidence that both the body shape in terms of the length to depth ratio and size at first maturity are under genetic control (Hagen, 1967; McPhail, 1977) and so are susceptible to selection. Resident lake fish tend to have a more streamlined body than stream-dwelling fish. In this respect, lake sticklebacks resemble anadromous populations more than stream populations, which again suggests convergence between lake and anadromous populations driven by similar selective factors (Hagen and Gilbertson, 1972).

Body size may be another character which is related to the presence of predators. The adult fish in Mayer Lake are unusually large. The mean standard length of sexually mature females and males was 89.94 and 80.24 mm, respectively, whereas in the streams the females had a mean length of 50.68 mm with males even smaller. Mayer Lake sticklebacks are predated by cut-throat trout and their unusually large size may allow them to swim faster and may also increase the minimum size of predators able to capture them, especially as their spines are also relatively long (Moodie, 1972a, b). A survey of 25 resident lake populations on Vancouver Island revealed significant inter-population variation in the size at which maturity was first reached, a variation that had a genetic basis. Although the selective factors that have produced these differences are not known, one possibility is predation. If large predators are rare, large body size may represent a size refuge from predation, so maturity should be delayed until that size is reached. If large predators are common, there is no size refuge so sexual maturity should be reached as soon as possible, which means at a small size (McPhail, 1977) (Chapter 7).

Variation in Breeding Colours

Throughout most of the geographical range of *G. aculeatus*, the male adopts breeding colours of a red throat, blue eyes and greenish-blue back (Chapter 7), but there is considerable variation in the development of these colours. In a few populations the deviation from the typical pattern is extensive. In two populations studied, the deviation is probably related to predation by trout, and in a third to the presence of a heterospecific competitor.

Only about 15 per cent of the males in Lake Wapato developed the typical red throat. The ventral surface of breeding males could be silver, drab with mottled black, completely black or red. The colour of the dorsal surface ranged from a mottled olive similar to the coloration of females, a silvery sky blue, a dark green to a jet black. There was no obvious correlation between the dorsal and ventral colorations. The frequency of males with red throats declined during the breeding season, although their nesting tended to occur in deeper water where it was more successful. Females showed a strong preference to mate with males with red throats whether the red was natural or painted on. Egg predation was common in the lake (Chapters 4 and 7) so it was expected that the males with red throats would be more successful at hatching their eggs given that the red throat intimidates rivals (Chapter 7). With these apparent advantages, why did all the males not have red throats since there would seem to be strong selection for the typical breeding pattern? The visually hunting rainbow trout probably found red males conspicuous and the presence of these predators countered any tendency for the proportion of red-throated males to increase (Semler, 1971). Laboratory experiments suggested that male sticklebacks with red throats were predated more by cut-throat trout (Moodie, 1972b).

The unusually large sticklebacks in Mayer Lake are melanic except that the opercula and antero-ventral region are bright silver. When sexually mature, the males lose the silver and assume a drab appearance. Most have sooty black throats, though 14-18 per cent do develop red throats. The blue eye and greenish back usually do not develop. Water in the lake is stained by peat, so the dark colours are likely to make the fish difficult to see by the predatory trout, an effect that would be reduced by a bright red throat (Moodie, 1972b).

A most intriguing series of populations in which the males have an unusual breeding coloration occur in some drainage systems of the Olympic Peninsula, Washington State (McPhail, 1969; Hagen and Moodie, 1979; Hagen *et al.*, 1980). The most important of these drainages is the Chehalis River. In these populations the males become jet black during the breeding season, but if they are disturbed, their colour fades rapidly and they become drab. The colour of breeding females is not unusual. These populations are surrounded, north and south, by populations with the characteristic male breeding colours and such typical populations also occur at some places within the drainage systems where they may be adjacent to populations with black males (Figure 10.5). Where populations are adjacent, there is

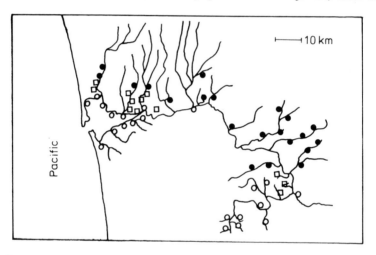

Figure 10.5: Distribution of Typical and 'Black' Populations of *G. aculeatus* in the Chehalis River System, Washington State. Open Circles, Typical Populations; Closed Circles, 'Black' Populations; Squares, Polymorphic Populations. Simplified after Hagen and Moodie (1979)

a steep cline between the typical and the 'black' populations although some hybridisation within a narrow zone does take place (Figure 10.6). Although 'black' populations occur throughout much of the Chehalis watershed, they are absent from the river mouths where the salinity is relatively high. A comparison of the salinity tolerances of typical and 'black' populations from two sites in the Chehalis system showed that the adults and eggs of the 'black' stickleback were less tolerant of high salinities than those from a typical population (McPhail, 1969).

The breeding colour is under simple genetic control, and hybrids

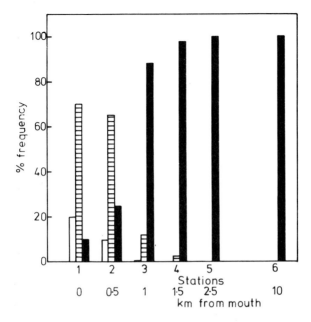

Figure 10.6: Frequency Distribution of 'Black' Sticklebacks in a Small Stream in Washington State. Black Bars, 'Black' Sticklebacks; Hatched Bars, Intermediates (Hybrids?); Open Bars, Typicals. Data from McPhail (1969)

between typicals and 'blacks' showed a mixture of black and red blotches on their throats. Laboratory experiments showed that there was a 5 per cent reduction in the viability of F_1 hybrids, which also showed a 5 per cent reduction in their fertility. Backcrosses between hybrids and parental forms indicated a 50 per cent reduction in the viability of the backcross zygotes. Hybrid males also showed unusually low rates of fanning their eggs, which resulted in high egg mortality. There is, then, a significant degree of reproductive isolation between the typical and 'black' populations, even when they inhabit the same stream, which results in a spatially restricted hybrid zone and little or no introgression.

When the mate choice of females from typical and 'black' populations was tested, both typical and 'black' females tended to choose the typical red males, unless the 'black' females came from a population which was adjacent to a typical population. In that case the 'black' female showed no significant preference for either male. This suggests that the unusual male coloration has evolved in spite of strong preferences by females for red-throated males. Why then have these unusual populations evolved?

An unusual feature of the drainage systems in which the 'black' populations are found is the presence of a small endemic freshwater fish, *Novumbra hubbsi*. The distribution of the 'black' sticklebacks coincides with the distribution of *N. hubbsi*. Where the latter is absent, the sticklebacks develop the typical male colours. This spatial congruence strongly suggests that the presence of *N. hubbsi* is the selective factor which has led to the evolution of the 'black' populations, but how?

Novumbra hubbsi is a carnivorous fish. When tested in the laboratory, it readily ate stickleback larvae, but the larvae from the 'black' population were more successful at evading predation than larvae from a typical population. There was also experimental evidence that *N. hubbsi* showed a tendency to orientate towards a nest guarded by a typical male rather than one guarded by a 'black' male. These observations led to the hypothesis that *N. hubbsi* was a predator on the larvae and perhaps the eggs of nesting sticklebacks, but the black colour of the unusual males and their ability to fade rapidly made them less conspicuous when guarding nests so that their nest contents suffered less predation than those of typical males. Thus predation by *N. hubbsi* was seen as the major selective factor favouring black breeding colours (McPhail, 1969).

Two observations showed that this hypothesis is probably wrong. Detailed sampling of *N. hubbsi* from natural populations found no evidence that they were predators on stickleback larvae: they ate invertebrates. Secondly, during their breeding season, which starts earlier but overlaps with that of the stickleback, male *N. hubbsi* adopt a dark-brown or black breeding colour and defend a territory. During territorial defence they will direct aggression towards intruding sticklebacks. Both the breeding colours and behaviour of male *N. hubbsi* show a striking similarity to that of the 'black' stickleback. In experiments a male *N. hubbsi* tended to face away from a black stickleback, which suggests it is intimidated by the latter's nuptial colour. The experiments showed that the orientation of *N. hubbsi* towards the nest guarded by a typical male described by McPhail (1969) was at least partly caused by a definite orientation away from a nest guarded by a 'black' male.

In the presence of a territorial *N. hubbsi*, male 'black' sticklebacks suffered fewer territorial intrusions by the *N. hubbsi* than typical males and were also more successful at nesting and hatching eggs. In contrast, in the absence of the *N. hubbsi*, typical males tended to hold larger territories and suffered fewer territorial intrusions by other sticklebacks than 'black' males. These results suggests that there is strong interspecific competition for territorial space between *N. hubbsi* and the sticklebacks. Under these conditions, the 'black' males are reproductively more successful than the typical males because they share the black throat coloration of the territorial *N. hubbsi*. The 'black' sticklebacks have converged on the black nuptial coloration of the endemic *N. hubbsi* (Hagen and Moodie, 1979; Hagen *et al.*, 1980). The sharp cline between typical and 'black' populations coinciding with the appearance of *N. hubbsi* suggests that there is strong selection against the 'black' stickleback outside the range of sympatry with *N. hubbsi*. This may be due to the preference of females for typical males but requires further study.

Other *G. aculeatus* populations in the Pacific North-West contain a small proportion of males in which black is an element in their nuptial coloration, which suggests that the genetic basis for rapid selection of the 'black' populations was present in the area (McPhail, 1969). A population containing some melanic males is found at the unusually high altitude of 1960 m in the San Bernadino Mountains in California, but the origin and significance of this population is not known (Bell, 1982b). Curiously, there are as yet no reports of populations with unusual male breeding colours from the European distribution.

Physiological and Biochemical Variation

Anadromous populations differ from resident populations in their salinity and temperature tolerances and in their ability to osmoregulate over a range of temperatures (Chapters 2 and 5). These differences are clearly adaptive and relate directly to the life-history patterns of these populations. Less attention has been paid to variations within populations. When fish from a Belgian resident population of lows were transferred directly from fresh water to full sea water (salinity 35 ppt) at 10°C, the mean length of survival was higher in the fish that had more plates. Fish with three or four plates survived for about 7.5 h whereas fish with seven or more plates survived, on average, for 13.7 h. In an equally drastic experiment, lows were transferred from freshwater just above 0°C to water at 25-28°C but with no change in salinity. In this case the fish with the lowest number of plates tended to survive longest (Heuts, 1947). These experiments suggest some correlation between plate phenotype and physiological tolerances, but this study has not been extended to more realistic environmental changes.

Over the last 20 years, the study of variation within natural populations has been revolutionised by the use of electrophoretic methods to detect protein variation (Ferguson, 1980). Surprisingly, this technique has only recently been used to study genetic variation within and between populations of sticklebacks. In the population at Friant, California, which is dimorphic for completes and lows, Avise (1976) found that of 15 genetic loci studied, 27 per cent were polymorphic, that is, had more than one allele. The mean percentage of individuals that were heterozygous per locus was 9.0 per cent for the lows and 7.5 per cent for the completes. These levels of polymorphism and heterozygosity are comparable to mean values for all vertebrates. In the population of lows in the Ventura drainage, also in California, less genetic variation was detected. Of eleven loci studied, only one (8.33 per cent) was polymorphic and the mean level of heterozygosity was 4.1 per cent (Bell and Richkind, 1981). A comparison of anadromous populations with resident populations in the Pacific North-West showed that the former tended to be genetically homogeneous over a relatively wide geographical area whereas the latter were genetically heterogeneous, even within small geographical areas (Withler, 1980, cited in McPhail and Hay, 1983). This parallels the situation for visible polymorphisms. Two mechanisms may contribute to this difference. In the anadromous populations, because of their virtually continuous coastal distribution during the winter, there is the possibility of migration between populations, with a resultant gene flow between populations maintaining a homogeneous gene pool. Migration between resident populations is less likely, partly because of physical barriers and partly because such populations seem to be relatively sedentary (Hagen, 1967). Secondly, selective factors may be more constant and similar over a relatively wide geographical range for anadromous populations (Gross, 1977), whereas there is now strong evidence that the selective factors can change within a few hundred metres in freshwater habitats. This point is well illustrated by the stickleback populations in and around the Mayer Lake and Chehalis River systems described previously.

Variation in Other Sticklebacks

Although the variation in *G. aculeatus* has attracted most attention, the other species are variable and in some cases show forms of variation that are comparable to those in *G. aculeatus*. This parallelism between species is reminiscent of the parallelism shown within *G. aculeatus*. In *P. pungitius* there is variation in the development of the lateral plates and in the number of dorsal spines and gill rakers, and in some populations spine-deficient phenotypes are found (Wootton, 1976; Gross, 1979). Differences in male breeding coloration, nest-site selection and reproductive behaviour also occur (Foster, 1977) (Chapter 7). Populations of *C. inconstans* from the north-west of its range frequently contain significant proportions of morphs showing a reduction or absence of the pelvic skeleton. Experiments and field observations have suggested that such spineless morphs show patterns of behaviour in the presence of a pike and some invertebrate predators that make them less likely to be predated than the morph with pelvic spines (Reist, 1980a,b, 1983). Breeding experiments showed that the degree of expression of the pelvic skeleton is under genetic control (Nelson, 1977). *Culaea inconstans* also shows significant inter-population variation in spine length, body shape, number of gill rakers and lateral plates (Moodie, 1977). There is significant intra- and inter-population variation in the number of dorsal spines in *A. quadracus*. The number can vary from one to seven, but most populations are polymorphic for four and five spines. Breeding experiments showed that this character had a high heritability and would respond rapidly to directional selection. This suggests that the observed frequencies of dorsal spine phenotypes in *A. quadracus* are maintained by selective factors (Blouw and Hagen, 1981; Hagen and Blouw, 1983).

Conclusion

The sticklebacks and *G. aculeatus* in particular show a paradoxical pattern of variation. There is an extremely wide geographical distribution in *G. aculeatus* throughout which typical populations of anadromous completes and resident lows occur. Although there is some geographical variation within these two groups, a traveller between the Pacific coast of North America and the Atlantic coast of Europe has no sense of any great differences. Yet populations occur which in characters such as pelvic skeleton development or male breeding coloration fall entirely outside the normal range of variation of the species. Furthermore, such populations with very similar phenotypes can be found separated by thousands of miles, for example the spine-deficient phenotypes in Scotland and British Columbia. It is also likely that resident populations of lows have evolved independently from anadromous populations of completes on several occasions. This evolutionary plasticity is constrained within tight limits for the stickle-backs have not shown any wide adaptive radiation such as that shown by the cichlid fishes (Fryer and Iles, 1972; Bell, 1976). This plasticity does provide almost unique material for the study of evolutionary change within vertebrates.

It is now feasible to compare different phenotypes in terms of their biochemistry, physiology, growth, feeding behaviour, reproductive success and survivorship. The generation length of sticklebacks is also sufficiently short to allow genetic analysis of the characters. Thus both the ecological and genetic components which maintain the unique pattern of variation shown by sticklebacks are open to study.

11 LIFE-HISTORY STRATEGY IN STICKLEBACKS

Introduction

Evolution has been likened to an existential game which the players cannot, in the long term, win. Failure is indeed ultimately inevitable, but success is measured by the length of time for which a player can avoid losing (Slobodkin and Rapoport, 1974). Individuals die, populations disappear, species go extinct, but if some individuals in a population are temporarily successful in their evolutionary game, they leave offspring that reach sexual maturity and so the population is maintained for at least another generation. Since the individuals that are successful may not be genetically fully representative of the whole population, the genetic composition of the population may change from one generation to the next. Evolution has taken place, but the continuity of the gene pool in time has not been broken.

At the present time, the sticklebacks are a successful family, and particularly the two species, *G. aculeatus* and *P. pungitius*. This success reflects the ability of large numbers of individual sticklebacks to produce offspring that in their turn become sexually mature. Why is the pattern of survivorship and reproduction shown by the sticklebacks successful? Over the past two decades, three major theories have been developed that attempt to predict what patterns of survivorship and reproduction will evolve in particular environments. These theories assume that life-history traits such as size and age at maturity, age-specific fecundity and survivorship are subject to natural selection, which tends to maximise the mean fitness of the population. The three theories are 'r- and K-selection', 'bet-hedging' and 'the cost of reproduction' (Stearns, 1976, 1977; Schaffer, 1979; Bell, 1980). How well do the life-history patterns of the sticklebacks conform to the assumptions and predictions of these models?

227

Theories of Life-history Evolution

r- and K-selection

Until recently, the theory of *r*- and *K*-selection was the most influential of the three. It is based on the logistic model of population growth, which can be expressed as:

$$dN/dt = rN\,(1 - N/K)$$

where N is the population density, r is the *per capita* rate of increase when the population is small, and K is the maximum density that can be sustained under the prevailing environmental conditions (Horn, 1978). The theory contrasts the life histories to be expected in two different types of environment. The first environment is harsh and unpredictable, so populations regularly suffer high rates of mortality caused by environmental factors whose effects on a population are independent of its density. A sparse population suffers as proportionately high losses as a dense one. Under these conditions, populations will rarely, if ever, reach a density of K. This environment will select for animals that have high potential rates of increase, which will result in populations recovering rapidly from catastrophic losses: evolution favours productivity. High rates of increase can be achieved by short generation times and high fecundities. The theory argues that the selection regimen favours early age at first reproduction, large clutch size, semelparity, a lack of parental care, numerous small offspring and a short generation time. In contrast, the second environment is physically benign, so a population can increase to a density at which competition for available resources such as food, space, and breeding sites becomes important. Both mortality and fecundity become density dependent, so at high population densities there is an increase in the mortality rate and a decrease in fecundity. Under these conditions, selection favours those animals that are able to breed successfully when in competition with other animals: evolution favours efficiency of conversion of food into progeny. The offspring must be of a quality that allows them to survive and reproduce in this competitive environment. The theory suggests that this situation selects for delayed reproduction, iteroparity, small clutches, parental care, and few but large offspring (Stearns, 1976).

It is tempting to use *r*- and *K*-selection as convenient labels for particular syndromes of life-history traits, but the concepts are more interesting in the context of a model that attempts to predict adaptive life-histories given specified environmental characteristics.

Bet-hedging

This theory assumes that an effective way of playing the evolutionary game is for an individual to attempt to minimise its chance of leaving no offspring at all. This theory also argues that the crucial factor determining the best life history is the variability of the environment. But, in contrast to the *r*- and *K*-selection model, it distinguishes between the situation where the environmental variability affects primarily juvenile survivorship and that in which adult survivorship is primarily affected. In the former, selection will favour spreading the risk of leaving no offspring over several breeding attempts. This will lead to a life history in which the animal is long-lived and iteroparous with a relatively low fecundity. In this situation, because adult survival is more certain than juvenile survival, adults are more valuable than juveniles. In the second situation, juvenile survival is more certain than that of adults: juveniles are more valuable. This will select for an early age at maturity and high fecundities but a short lifespan (Stearns, 1976).

Costs of Reproduction

The two previous theories, although they have been useful in stimulating both theoretical and empirical research, have a fundamental flaw. They do not suggest why an animal cannot mature at an early age and have a high fecundity yet still have a long lifespan and repeatedly exhibit high fecundity. They both imply that early maturation and high fecundity are inversely correlated with a long lifespan, but do not indicate why this pattern of covariation of life-history traits occurs. The third theory deals explicitly with this point. It assumes that any current breeding attempt imposes a cost such that future expected reproductive success is reduced. There is a trade-off between current reproduction and future expected reproduction. Since the total lifetime reproductive output of an individual is the sum of its present reproductive output and expected future outputs, selection should favour a life-history pattern which at each moment in time tends to maximise this sum. There is a complication that young produced relatively early in life can be more valuable than young produced late in life because the former will probably become sexually mature and start producing their own offspring well before the later offspring. This will tend to favour current reproduction over future reproduction. In fish, there is another complication, because of their indeterminate growth pattern. Growth continues after sexual maturity, and with increase in size there is usually an increase in fecundity, especially in females, but also in males if large size gives an advantage in access to females. If a current reproduction attempt involves a use of resources that could have been used for growth, then by reducing growth it will reduce future expected reproduction even if the current reproductive event has no effect on subsequent survival.

The analysis of this model has been pioneered by Schaffer (1974), Schaffer and Rosenzweig (1977) and Bell (1980). Their analyses suggest that a crucial factor in determining which is the best life history is the shape, for each age class, of the trade-off curve between current reproductive output and future expected reproduction. If the trade-off curve is concave, then fitness can be maximised by either reproduction without post-breeding survival, that is semelparity, or no reproduction but high survival to the next age class. The level of future expected reproduction will determine which of these is adopted. A convex trade-off curve leads to an iteroparous life history, because particular combinations of current reproductive output and post-breeding survival and growth will maximise the lifetime production of offspring. In this case, the balance between the two may be determined by the chances of the juveniles surviving to reach the age at maturity. If juvenile survival is good, a high current reproductive output and a poor post-breeding survival is favoured, whereas if juvenile survival is poor, a low current reproductive output and good post-breeding survival is favoured (Schaffer and Rosenzweig, 1977; Calow and Sibly, 1983). This prediction is identical to that of the bet-hedging model.

A test of the reproduction-cost model involves showing that current reproduction does exert a cost on future expected reproduction because of its effect on survivorship, growth or a combination of the two. Ideally, it should also be tested by determining the shape of the trade-off curve for animals of a known life-history pattern.

Reproductive Effort

Some insight into the trade-off between current reproduction and future expected reproduction can be obtained through the use of the concept of reproductive effort. This is defined as the proportion of the total resources available that is allocated to reproduction over some defined time period (Gadgil and Bossert, 1970). The two crucial resources are food, which provides the energy and nutrients that can be spent on reproduction, and time. For an animal, both of these resources are finite, so expenditure of a resource on one activity reduces the supply of the resource that is available to be spent on other activities. The model of energy partitioning described in Chapter 5 indicates that the energy income of the fish must be partitioned between several expenditures, so that energy expended on producing gametes or on reproductive behaviour will not be available for somatic growth and activities related to foraging or escape from predators. Similarly, time that is spent on reproductive activities such as courtship, nest building, territorial defence and parental behaviour means that less time is available for searching for food or hiding from predators.

A high reproductive effort will be correlated with a high current reproductive output but a reduction in future expected reproduction, because of a reduction in post-breeding survivorship and growth. In the *r*- and *K*-selection model, it is assumed that *r*-selection leads to a high reproductive effort at an early age of maturity, whereas *K*-selection results in a relatively low reproductive effort at each breeding attempt but good adult survivorship and perhaps good growth. The bet-hedging model assumes that if juvenile survival is poor or variable, this will select for a low reproductive effort at each breeding attempt, whereas poor or variable adult survivorship will select for a high reproductive effort. In the reproduction-cost model, the shape of the trade-off curve can be interpreted as a result of the shape of the relationships between reproductive effort and current reproduction, and future expected reproduction (Schaffer, 1979).

Life-history Strategies of the Sticklebacks

Although there are many variations in the details of the life histories of the sticklebacks, the main themes are common to all the species. They are small fish that become reproductively mature at a relatively early age and they have a short lifespan. Age at maturity and lifespan show both intra- and inter-specific variation. In *G. aculeatus*, there is slight evidence that the age at maturity and the lifespan are greater in more northerly populations, but this needs to be confirmed by life-history studies over a wide geographical range. Although in many populations few if any fish survive to a second breeding season, the sticklebacks do not seem to include any species that have populations that are obligately semelparous. More detailed studies of *S. spinachia* may alter this generalisation. In addition to their potential of being iteroparous over breeding seasons, the sticklebacks are iteroparous within a breeding season. Females have the potential of producing clutches several times in a breeding season, the number of clutches produced depending strongly on the food supplies. Males also have the potential for nesting and rearing young more than once per season. Reproduction in the sticklebacks is characterised by male territoriality and nest building and elaborate courtship displays. Males are also parental: eggs will not develop without the attentions of the male parent. In some species, there are populations that make a spawning migration from coastal waters on to spawning grounds in estuaries and rivers. Such migratory populations may coexist in the same river systems as populations that spend their whole lives within the system.

How well does this combination of life-history traits conform to the expectations of the theories of life histories outlined above?

A Critique of the Theories

r- and K-selection. The combination of life-history traits that characterises the sticklebacks fits very uneasily into the scheme predicted by the model of *r-* and *K*-selection. Some of the traits, such as small size, early age of maturity and short lifespan, might be regarded as typical of a *r*-selected animal. Other traits, especially the elaborate reproductive behaviour, which includes parental care, are more characteristic of *K*-selected animals. The eggs of the sticklebacks are not unusually small; rather, their size is typical of summer-spawning freshwater teleosts from the north temperate zone (Wootton, 1984), so sticklebacks do not increase their fecundity by producing small propagules. Their fecundity is enhanced by their ability to spawn several times within a breeding season.

A further problem is that the geographical range and variety of habitats that the sticklebacks, especially *G. aculeatus* and *P. pungitius*, inhabit make it impossible to categorise their typical environment as either harsh or benign. Rather, populations are found in environments that range from the relatively harsh, such as the arctic and sub-arctic, through to the relatively benign environments in the central and southern parts of the geographical distribution. In addition, the ability of the sticklebacks to tolerate wide ranges of salinity and temperature (Chapter 5) reduces the chances of the population being subjected to catastrophic losses caused by abiotic factors.

The model of *r-* and *K*-selection does not have built into it sufficient detail to account for the covariation in life-history traits that is observed in natural populations. High values of *r* or good competitive abilities when the population abundance is close to *K* cannot simply be equated with either high birth rates or good survivorship, respectively (Schaffer, 1979), nor does the model incorporate details of age-specific fecundities and mortalities which have major implications for life-history pattern (Charlesworth, 1980). In its present form, this model of the evolution of life-history traits must be regarded as obsolete and its inability to comprehend the details of the life history of the sticklebacks merely emphasises this.

Bet-hedging. Chapter 9 noted that the demography of stickleback populations has received insufficient study. Age-specific mortality rates are not known, so it is impossible to compare the juvenile and adult survivorships in a way that would allow the bet-hedging model to be adequately tested. The life-history traits of the sticklebacks are those that would be expected of populations in which juvenile survival was relatively good or less variable than adult survival. Given the high level of parental care, which would be expected to enhance juvenile survival perhaps at the cost of a reduction in the survivorship of the parents, the predictions of the model are plausible. Juvenile sticklebacks quickly start to exploit the same range of food items as their parents, but their absolute requirements in the first few months of life are considerably less than those of the adults because of the difference in body size (Chapter 4). A rate of exploitation of the food supply which left insufficient for the adults could still provide enough for the juveniles, so that again adult survivorship would be poor compared with that of the juveniles. A comparison of the life-history traits found in populations in which juvenile and adult survivorships were measured would provide a useful test of how well the model predicted the evolution of these traits in sticklebacks.

Reproduction Cost Model. In his classic *Adaptation and Natural Selection*, Williams (1966) listed a number of life-history traits which he thought were correlated with an increased reproductive effort. They included: a large ratio of progeny weight to female body weight; several clutches per season; bright coloration in breeding males; lengthy and elaborate courtship display; territorial and aggressive behaviour; and parental care. This almost reads like a list of the life-history characteristics of the sticklebacks. Because of the ease with which *G. aculeatus* can be bred in the laboratory, the reproductive effort of the female can be measured in terms of the weight (or energy content) of eggs produced as a proportion of the weight (or energy content) of food consumed (Wootton and Evans, 1976; Wootton, 1977; D.A. Fletcher and R.J. Wootton, unpublished). This reproductive effort can be estimated either for the interval between successive spawnings within a breeding season or for the whole breeding season. Because the male's reproductive effort is mainly in terms of behavioural invest-ment, it has proved more difficult to measure and cannot yet be con-sidered. For the male, it may be better to measure reproductive effort in terms of expenditure of time rather than energy.

Experiments with female *G. aculeatus* have yielded estimates of reproductive effort and also tested some of the basic assumptions of the reproduction-cost model. Females fed *Tubifex* ad lib had a mean repro-ductive effort of 26 per cent with a range from 9 to 42 per cent. Effort was estimated by measuring the total energy content of the eggs pro-duced at a spawning as a percentage of the energy content of the food consumed in the time between that spawning and the previous one, that is, the food consumed over the inter-spawning interval (Wootton and Evans, 1976). When females were fed enchytraeid worms, the reproductive effort per inter-spawning interval ranged from 5.3 per cent to 39.9 per cent, and was inversely correlated with the ration size. At a ration size at which there was no net change in the weight of a female over an inter-spawning interval, reproductive effort was higher in heavier fish. On average, a female weighing 0.8 g had a reproductive effort of 15.8 per cent, but a female weighing 1.6 g had an effort of 21.4 per cent (Wootton, 1977). Over a breeding season, females fed enchytraeid worms still showed an inverse correlation between repro-ductive effort and ration size (Table 11.1). Over both a single inter-spawning interval and an entire breeding season, at a given ration level there was a positive correlation between fecundity and reproductive effort. In contrast, there was a negative correlation between growth rate and reproductive effort. At low rations, females lost weight during the breeding season, even at rations which outside this season were sufficient to allow some growth. These correlations are in the directions assumed by the reproduction-cost model. There was no significant correlation between mortality during or immediately after the breeding season and reproductive effort, but because of the relatively small numbers of females used, the differences in mortality would have had to be large to be detected (D.A. Fletcher and R.J. Wootton, unpublished).

The results of these experiments are compatible with the reproduction-cost model, but the variation in the data was too great to indicate the shape of the trade-off curves. The results are also comparable to those from similar experiments on the small cyprinodont teleost, *Oryzias latipes* (Hirshfield, 1980). In *O. latipes*, the levels of reproductive effort were relatively similar to those shown by *G. aculeatus*, and there

Table 11.1: Relationship between Nominal Ration Level and
Reproductive Effort[a] Over a Breeding Season for Female *G. aculeatus*

Number of fish	Ration as % bodyweight	Reproductive effort (with standard deviation)
7	2	39.9 ± 29.9
10	4	27.0 ± 18.1
11	8	16.2 ± 5.9
8	16	15.4 ± 4.5

[a] Reproductive effort defined as weight of eggs produced during breeding season as a percentage of weight of food consumed over the same period.

was also a tendency for the effort to be inversely related to ration. Additionally, there was a significant positive correlation between mortality and reproductive effort. Like the sticklebacks, *O. latipes* is a small, short-lived species with an early age of maturation. Measurements of the reproductive efforts of longer-lived, larger, iteroparous species are needed for further comparative analyses of this aspect of the reproduction-cost model to be made.

Although comparable experiments have not yet been conducted on the male stickleback, there are several grounds for predicting that reproduction will impose a cost on both growth and survival. Territorial and parental behaviour will tend to restrict the foraging of the males both spatially and temporally at a time when activities such as territorial defence, nest care and fanning the developing eggs will represent significant expenditures of energy. Time spent in these reproductive behaviours will also allow less time to be spent in hiding from or evading predators, and such behaviours make the male more conspicuous (Giles, 1984). Territorial fighting may cause damage to the skin and facilitate parasitic and bacterial infections. There are so many anecdotal reports of high post-breeding mortality in sticklebacks that a mortality cost associated with reproduction seems certain.

None of the experimental results or observational details are incompatible with the reproduction-cost model, and the scope for both intra- and inter-specific comparisons of experimental and field studies is such that sticklebacks potentially provide a valuable test case for that model.

Evolution of Small Size

The cause-and-effect relationships in the evolution of life-history characteristics are difficult, if not impossible, to tease apart. The relatively small size of the sticklebacks is one such characteristic. Has the small body size evolved in response to demographic selection factors which favour an early age of sexual maturity over a long period of juvenile growth, or have other factors selected for a small body size and the demographic characteristics co-evolved to accommodate this evolutionary tendency? Evolution of small body size is not rare in the teleosts: probably 10 per cent of all teleostean species have a maximum adult length of less than 100 mm (Lindsey, 1966; Miller, 1979b). Although the proportion of small species tends to increase at lower latitudes, they still form a significant portion of the teleost fauna at high latitudes, especially in fresh waters. What are the advantages of small size?

The theories of life-history strategies discussed earlier suggest that if the mortality of adults is relatively greater or more variable than that of juveniles, then this demographic pattern will select for early breeding with a high reproductive effort. As a consequence, sexual maturity will be reached at a small size and post-breeding survivorship will be reduced. This will in turn reinforce the selection for early maturation, and may select for improved juvenile survivorship through the evolution of some form of parental care. The factors that initially led to the increase in adult mortality will probably remain unknown, because of the time that has elapsed since the evolutionary process began.

An advantage of small size is that sexual maturity can be reached even when the environment is chronically short of food, or the available food consists mainly of small food items such as small zooplankters which are spatially too dispersed to be successfully exploited by filter feeders but can be exploited by small fish feeding on individual items. If such feeding conditions prevail, survivorship of large adults may be compromised, which would add to the selection pressures in favour of small body size. Small size will also allow the fish to exploit a topographically complex substrate or environments that are spatially restricted (Miller, 1979b). The evolution of small size to exploit such environments would entail concomitant changes in the life-history characteristics. Small size will usually mean a reduced fecundity and an increased risk of predation because it will make the fish the potential prey of a larger range of predators, including large invertebrates (Chapter 8).

Evolution of small body size may also occur if the important predators are size selective. If there is a minimum size of prey below which it is not profitable for the predators to take them, prey that reach sexual maturity at a size less than that taken by the predators may have a selective advantage over prey that become exposed to predation at or before the time they become sexually mature (McPhail, 1977).

It is not known which, if any, of these processes has led to the evolution of small size in the sticklebacks. The fossil record of *Gasterosteus* stretches back to the Miocene (Bell, 1977), but the earlier evolutionary history of the family is obscure.

Migratory Behaviour

The modern distribution and physiological tolerances of the sticklebacks indicate they are essentially a marine family that has successfully invaded fresh water. In some populations this is reflected in anadromous behaviour, the fish spawning in fresh or brackish water but moving to coastal waters to overwinter. Such anadromous populations can coexist in the same river system with populations that do not migrate, so an anadromous migration is not obligatory. A feature of the fish of anadromous populations is that they tend to reach a larger mean size than resident populations and partly as a consequence they have a higher mean fecundity (Figure 7.7) (Hagen, 1967). If the increased fecundity is sufficient to balance the poorer survivorship that is probably an inevitable consequence of a migration, then the potential rate of increase of the anadromous population would equal that of the resident population: the two would have equal mean fitnesses. In practice, the anadromous and resident populations also tend to live in different parts of the river system, which makes them less direct competitors and reduces the gene flow between them (Chapter 10). The evolutionary history of the sticklebacks seems to be of repeated and successful colonisations of the freshwater habitat from a marine base (Bell, 1976), and the anadromous habit of some populations must have facilitated this exploitation of fresh waters. It is possible that the rates of predation on eggs and young fish are lower in fresh waters, which would result in selection for breeding and spending at least the early period of life in fresh water. This advantage may be partly offset by poorer food availability in fresh water and problems of osmoregulation in a hyposmotic environment.

Conclusion

Sticklebacks provide an almost unique material for the study of the evolution of life-history patterns. There is scope for inter-specific comparisons, which is now being exploited (FitzGerald, 1983; Rowland, 1983a,b). The wide geographical and habitat ranges of *G. aculeatus* and *P. pungitius* provide opportunities for inter-population comparisons of life-history characteristics. All stages of the life history can be studied in the laboratory and in the field with relative ease. Genetic analyses of life-history traits are feasible because of the ease with which sticklebacks can be bred and their short life history. Much valuable information can be obtained without demanding expensive apparatus and technical skill, so life-history studies can form the basis of field exercises and laboratory projects for undergraduates and even schoolchildren. I hope that this book will stimulate such projects, and that it will encourage communication between devotees of the biology of the sticklebacks so that some of the problems posed by their diversity of life histories can be solved.

REFERENCES

Abdel'-Malek, S.A. (1968). 'Feeding of Young Three-Spined Stickle-backs (*Gasterosteus aculeatus* L.) in Kandalaksha Bay in the White Sea.' *Probl. Ichthyol. 8*, 74-80

Ahokas, R.A. and Duerr, F.G. (1975). 'Salinity Tolerance and Extra-cellular Osmoregulation in Two Species of Euryhaline Teleosts, *Culaea inconstans* and *Fundulus diaphanus*.' *Comp. Biochem. Physiol. 52A*, 445-8

Allen, J.R.M. (1980). 'The Estimation of Natural Feeding Rate of the Threespined Stickleback (*Gasterosteus aculeatus* L.) (Pisces),.' PhD Thesis, University of Wales

Allen, J.R.M. and Wootton, R.J. (1982a). 'The Effect of Ration and Temperature on the Growth of the Three-spined Stickleback, *Gasterosteus aculeatus* L.' *J. Fish Biol. 20*, 409-22

Allen, J.R.M. and Wootton, R.J. (1982b). 'Age, Growth and Rate of Food Consumption in an Upland Population of the Three-spined Stickleback, *Gasterosteus aculeatus* L.' *J. Fish Biol. 21*, 95-105

Allen, J.R.M. and Wootton, R.J. (1982c). 'Effect of Food on the Growth of Carcase, Liver and Ovary in female *Gasterosteus aculeatus* L.' *J. Fish Biol. 21*, 537-47

Allen, J.R.M. and Wootton, R.J. (1983). 'Rate of Food Consumption in a Population of Threespine Sticklebacks, *Gasterosteus aculeatus*, Estimated from the Faecal Production.' *Env. Biol. Fish. 8*, 157-62

Allen, J.R.M. and Wootton, R.J. (1984). 'Temporal Patterns in Diet and Rate of Food Consumption of the Three-spined Stickleback (*Gasterosteus aculeatus* L.) in Llyn Frongoch, an Upland Welsh Lake.' *Freshwat. Biol.* in press

Anderson, R.M. (1981). 'Population Ecology of Infective Disease Agents.' In *Theoretical Ecology* (R.M. May, ed.) pp. 318-55. Black-well, Oxford

Aneer, G. (1973). 'Biometrical Characteristics of the Three-spined Stickleback (*Gasterosteus aculeatus* L.) from Northern Baltic Proper.' *Zool. Scr. 2*, 157-62

Anker, G. Ch. (1978). 'Analysis of Respiration and Feeding Movements of the Three-spined Stickleback, *Gasterosteus aculeatus* L.' *Neth. J. Zool. 28*, 485-523

Arme, C. and Owen, R.W. (1967). 'Infections of the Three-spined Stickleback, *Gasterosteus aculeatus* L., with the Plerocercoid Larvae of *Schistocephalus solidus* (Muller, 1776), with Special Reference to Pathological Effects.' *Parasitology 57*, 301-14

Assem, J. van den (1967). 'Territory in the Three-spined Stickleback, *Gasterosteus aculeatus* L. An Experimental Study in Intra-specific Competition.' *Behaviour Suppl. 16*, 1-164

Assem, J. van den and Molen, J.N. van der (1969). 'Waning of the Aggressive Response in the Three-spined Stickleback upon Constant Exposure to a Conspecific. I. A Preliminary Analysis of the Phenomenon.' *Behaviour 36*, 286-334

Avise, J.C. (1976). 'Genetics of Plate Morphology in an Unusual Population of Threespine Sticklebacks (*Gasterosteus aculeatus*).' *Genet. Res. 27*, 33-46

Baggerman, B. (1957). 'An Experimental Study on the Timing of Breeding and Migration in the Three-spined Stickleback.' *Arch. neerl. Zool. 12*, 105-317

Baggerman, B. (1968). 'Hormonal Control of Reproductive and Parental Behaviour in Fishes.' In *Perspectives in Endocrinology: Hormones in the Lives of Lower Vertebrates* (E.J.W. Barrington and C. Barker Jorgensen, eds) pp. 351-404. Academic Press, London

Baggerman, B. (1980). 'Photoperiodic and Endogenous Control of the Annual Reproductive Cycle in Teleost Fishes.' In *Environmental Physiology of Fishes* (M.A. Ali, ed.) pp. 533-567. Plenum, New York

Baker, M.C. (1971). 'Habitat Selection in Fourspine Sticklebacks (*Apeltes quadracus*).' *Amer. Midl. Nat. 85*, 239-42

Bakker, Th. C.M. and Sevenster, P. (1983). 'Determinants of Dominance in Male Sticklebacks (*Gasterosteus aculeatus* L.).' *Behaviour 86*, 55-71

Bayer, R.D. (1980). 'Size and Age of the Tube-snout (*Aulorhynchus flavidus*) in the Yquina Estuary, Oregon.' *Northwest. Nat. 54*, 306-10

Begon, M. (1979). *Investigating Animal Abundance*. Edward Arnold, London

Bell, G. (1980). 'The Costs of Reproduction and their Consequences.' *Amer. Nat. 116*, 45-76

Bell, M.A. (1973). 'The Pliocene Stickleback, *Pungitius haynesi*, a Junior Synonym of *Gasterosteus aculeatus*.' *Copeia* 1973, 588-90

Bell, M.A. (1974). 'Reduction and Loss of the Pelvic Girdle in *Gasterosteus* (Pisces): a Case of Parallel Evolution.' *Contrib. Sci. Nat. Hist. Mus. Los Angeles Cty 257*, 1-36

Bell, M.A. (1976). 'Evolution of Phenotypic Diversity in *Gasterosteus aculeatus* Superspecies on the Pacific Coast of North America.' *Syst. Zool. 25*, 211-27

Bell, M.A. (1977). 'A Late Miocene Marine Threespine Stickleback, *Gasterosteus aculeatus*, and its Zoogeographical and Evolutionary Significance.' *Copeia* 1977, 277-81

Bell, M.A. (1981). 'Lateral Plate Polymorphism and Ontogeny of the Complete Plate Morph of Threespine Sticklebacks (*Gasterosteus aculeatus*).' *Evolution 35*, 67-74

Bell, M.A. (1982a). 'Differentiation of Adjacent Stream Populations of Threespine Sticklebacks.' *Evolution 36*, 189-99

Bell, M.A. (1982b). 'Melanism in a High Elevation Population of

Gasterosteus aculeatus.' Copeia 1982, 829-34

Bell, M.A. and Haglund, T.R. (1978). 'Selective Predation of Threespine Sticklebacks (*Gasterosteus aculeatus*) by Garter Snakes.' *Evolution 32*, 304-19

Bell, M.A. and Richkind, K.E. (1981). 'Clinal Variation of Lateral Plates in Threespine Stickleback Fish.' *Amer. Nat. 117*, 113-32

Bengtson, S. (1971). 'Food and Feeding of Diving Ducks at Lake Myvatn, Iceland.' *Ornis. Fenn. 48*, 77-92

Benjamin, M. (1974). 'Seasonal Changes in the Prolactin Cells of the Pituitary Gland of the Freshwater Stickleback, *Gasterosteus aculeatus*, Form *leiurus*.' *Cell Tiss. Res. 152*, 93-102

Benzie, V.L. (1965). 'Some Aspects of Anti-predator Responses of Two Species of Stickleback.' D Phil Thesis, University of Oxford

Bertin, L. (1925). 'Recherches Bionomique, Biometriques et Systematiques sur les Epinoches (Gasterosteides).' *Ann. Inst. Oceanogr. Monaco 2*, 1-204

Bertram, B.C.R. (1978). 'Living in Groups: Predators and Prey.' In *Behavioural Ecology: an Evolutionary Approach* (J.R. Krebs and N.B. Davies, eds) pp. 64-96. Blackwell, Oxford

Beukema, J.J. (1968). 'Predation by the Three-spined Stickleback (*Gasterosteus aculeatus*): the Influence of Hunger and Experience.' *Behaviour 31*, 1-126

Bianco, P.G. (1980). 'Areale Italico, Rinventimento in Calabria e Origini delle Popolazioni Mediterranee di *Gasterosteus aculeatus* L.' *Boll. Mus. Civ. St. Nat. Verona 7*, 197-216

Black, R. (1971). 'Hatching Success in the Three-spined Stickleback (*Gasterosteus aculeatus*) in Relation to Changes in Behaviour during the Parental Phase.' *Anim. Behav. 19*, 532-41

Black, R. and Wootton, R.J. (1970). 'Dispersion in a Natural Population of Three-spined Sticklebacks.' *Can. J. Zool. 48*, 1133-5

Blahm, T.H. and Snyder, G.R. (1975). 'Effect of Increased Water Temperatures on Survival of Adult Threespine Stickleback and Juvenile Yellow Perch in the Columbia River.' *Northwest. Sci 49*, 267-70

Blouw, D.M. and Hagen, D.W. (1981). 'Ecology of the Fourspine Stickleback, *Apeltes quadracus*, with Respect to a Polymorphism for Dorsal Spine Number.' *Can. J. Zool. 59*, 1677-92

Boddeke, R., Slijper, E.J. and Stelt, A. van der (1959). 'Histological Characteristics of the Body Musculature of Fishes in Connection with their Mode of Life.' *Proc. K. Ned. Akad. Wet. Ser. C 62*, 576-88

Borg, B. (1981). 'Effects of Methyltestosterone on Spermatogenesis and Secondary Sexual Characters in the Three-spined Stickleback (*Gasterosteus aculeatus* L.).' *Gen. Comp. Endocrinol. 44*, 177-80

Borg, B. (1982a). 'Seasonal Effects of Photoperiod and Temperature on Spermatogenesis and Male Secondary Sexual Characters in the Three-spined stickleback, *Gasterosteus aculeatus* L.' *Can. J. Zool. 60*, 3377-86

Borg, B. (1982b). 'Extraretinal Photoreception Involved in Photoperiodic Effects on Reproduction in Male Three-spined Sticklebacks

Gasterosteus aculeatus.' Gen. Comp. Endocrinol. 47, 84-7

Borg, B. and Ekstrom, P. (1981). 'Gonadal Effects of Melatonin in the Three-spined Stickleback, *Gasterosteus aculeatus* L. during Different Seasons and Photoperiods.' *Repr. Nutr. Dev. 21*, 919-28

Borg, B. and Veen, Th. van (1982). 'Seasonal Effects of Photoperiod and Temperature on the Ovary of the Three-spined Stickleback, *Gasterosteus aculeatus* L.' *Can. J. Zool. 60*, 3387-93

Borg, B., Goos, H.J.Th. and Terlou, M. (1982). 'LHRH-immunoreactive Cells in the Brain of the Three-spined Stickleback, *Gasterosteus aculeatus* L.' (Gasterosteidae). *Cell Tiss. Res. 226*, 695-9

Braekevelt, C.R. and McMillan, D.B. (1967). 'Cyclic Changes in the Ovary of the Brook Stickleback *Eucalia inconstans* (Kirtland).' *J. Morph. 123*, 373-96

Brett, J.R. (1979). 'Environmental Factors and Growth.' In *Fish Physiology*, Vol. VIII (W.S. Hoar, D.J. Randall and J.R. Brett, eds) pp. 599-675. Academic Press, London

Brown, W.W. and Cheng, C. (1946). 'Investigations into the Food of the Cod (*Gadus callarius* L.) off Bear Island, and of the Cod and Haddock (*G. aeglefinus* L.) off Iceland and the Murman Coast.' *Hull Bull. mar. Ecol. 3*, 35-71

Burton, D. (1975). 'The Integumentary Melanophore Patterns of Two Teleost Species *Gasterosteus aculeatus* and *Pseudopleuronectes americanus*.' *Can. J. Zool. 53*, 507-15

Burton, D. (1978). 'Melanophore Distribution within the Integumentary Tissue of Two Teleost Species, *Pseudopleuronectes americanus* and *Gasterosteus aculeatus* Form *leiurus*.' *Can. J. Zool. 56*, 526-35

Calow, P. (1973). 'The Relationship between Fecundity, Phenology and Longevity: a Systems Approach.' *Amer. Nat. 107*, 559-74

Calow, P. and Sibly, R.M. (1983). 'Physiological Trade-offs and the Evolution of Life Cycles.' *Sci. Prog., Oxf. 68*, 177-88

Cameron, J.N., Kostoris, J. and Penhale, P.A. (1973). 'Preliminary Energy Budget of the Nine-spine Stickleback (*Pungitius pungitius*) in an Arctic Lake.' *J. Fish. Res. Bd Can. 30*, 1179-89

Campbell, R.N. (1979). 'Sticklebacks (*Gasterosteus aculeatus* L. and *Pungitius pungitius* (L.)) in the Outer Hebrides, Scotland.' *Hebrid. Nat. 3*, 6-15

Charlesworth, B. (1980). *Evolution in Age-structured Populations.* Cambridge University Press, Cambridge

Clarke, B.C. (1962). 'Balanced Polymorphism and the Diversity of Sympatric Species.' *Syst. Assoc. Publ. 4*, 47-70

Coad, B.W. (1983). 'Plate Morphs in Freshwater Samples of *Gasterosteus aculeatus* from Arctic and Atlantic Canada: Complementary Comments on a Recent Contribution.' *Can. J. Zool. 61*, 1174-7

Coad, B.W. and Power, G. (1973a). 'Observations on the Ecology and Meristic Variation of the Ninespine Stickleback, *Pungitius pungitius* (L. 1758) of the Matamek River System, Quebec.' *Amer. Midl. Nat. 90*, 498-503

Coad, B.W. and Power, G. (1973b). 'Observations on the Ecology of Lacustrine Population of the Three-spine Stickleback (*Gasterosteus aculeatus* L. 1758) in Matamek River System, Quebec.' *Nat. Can.* 100, 437-45

Cohen, S. and McFarland, D. (1979). 'Time-sharing as a Mechanism for the Control of Behaviour Sequences during the Courtship of the Three-spined Stickleback (*Gasterosteus aculeatus*).' *Anim. Behav.* 27, 270-83

Cole, S.J. (1978). 'Studies on the Energy Budgets of Two Freshwater Teleosts.' MSc Thesis, University of Wales

Courtenay, S.C. and Keenleyside, M.H.A. (1983). 'Nest Site Selection by the Fourspine Stickleback, *Apeltes quadracus*.' *Can. J. Zool. 61*, 1443-7

Craig, D. and FitzGerald, G.J. (1982). 'Reproductive Tactics of Four Sympatric Sticklebacks (Gasterosteidae).' *Env. Biol. Fish. 7*, 369-75

Cronly-Dillon, J. and Sharma, S.C. (1968). 'Effect of Season and Sex on the Photopic Spectral Sensitivity of the Three-spined Stickleback.' *J. Exp. Biol. 49*, 679-87

Davies, N.B. (1978). 'Ecological Questions about Territorial Behaviour.' In *Behavioural Ecology: An Evolutionary Approach* (J.R. Krebs and N.B. Davies, eds) pp. 317-350. Blackwell, Oxford

De Bruin, J.P.C. (1980). 'Telencephalon and Behaviour in Teleost Fish. A Neuroethological Approach.' In *Comparative Neurology of the Telencephalon* (S.O.E. Ebbesson, ed.) pp. 175-201. Plenum, London

De Ruiter, A.J.H. (1980). 'Effects of Testosterone on Kidney Structure and Hydromineral Regulation in the Teleost *Gasterosteus aculeatus* L.' Doctoral Thesis, University of Groningen

De Ruiter, A.J.H. and Mein, C.G. (1982). 'Testosterone-dependent Transformation of Nephronic Tubule Cells into Serous and Mucus Gland Cells in Stickleback Kidneys *in vivo* and *in vitro*.' *Gen. Comp. Endocrinol. 47*, 70-83

Downhower, J.F. and Brown, L. (1981). 'The Timing of Reproduction and its Behavioural Consequences for Mottled Sculpins, *Cottus bairdi*. In *Natural Selection and Social Behaviour* (R.D. Alexander and D.W. Tinkle, eds) pp. 78-95. Chiron, New York

Edmunds, M. (1974). *Defence in Animals*. Longman, London

Eggers, D.M. (1977). 'The Nature of Prey Selection by Planktivorous Fish.' *Ecology 58*, 46-59

Eggers, D.M. (1979). 'Comments on Some Recent Methods for Estimating Food Consumption by Fish.' *J. Fish. Res. Bd Can. 36*, 1018-19

Eggers, D.M., Bartoo, N.W., Rickard, N.A., Nelson, R.E., Wissmar, R.C., Burgner, R.L. and Devol, A.H. (1978). 'The Lake Washington Ecosystem: the Perspective from the Fish Community Production and Forage Base.' *J. Fish. Res. Bd Can. 35*, 1553-71

Elliott, J.M. (1979). 'Energetics of Freshwater Teleosts.' *Symp. zool. Soc. Lond. 44*, 29-61

Elliott, J.M. (1981). 'Some Aspects of Thermal Stress on Freshwater Teleosts.' In *Stress and Fish* (A.D. Pickering, ed.) pp. 209-45. Academic Press, London

Elliott, J.M. and Davison, W. (1975). 'Energy Equivalents of Oxygen Consumption in Animal Energetics.' *Oecologia 19*, 195-201

Elliott, J.M. and Persson, L. (1978). 'The Estimation of Daily Rates of Food Consumption for Fish.' *J. Anim. Ecol. 47*, 977-91

Emery, A.R. (1973). 'Preliminary Comparisons of Day and Night Habits of Freshwater Fish in Ontario Lakes.' *J. Fish. Res. Bd Can. 30*, 761-74

Endler, J.A. (1977). *Geographic Variation, Speciation and Clines.* Princeton University Press, Princeton

Falconer, D.S. (1981). *Introduction to Quantitative Genetics*, 2nd edn. Longmans, London

Fange, R. (1953). 'The Mechanism of Gas Transport in the Euphysoclist Swimbladder.' *Acta Physiol. Scand. 23*, Suppl. 110, 1-133

Feldmeth, R.C. and Baskin, J.N. (1976). 'Thermal and Respiratory Studies with Reference to Temperature and Oxygen Tolerance for the Unarmoured Stickleback *Gasterosteus aculeatus williamsoni* Hubbs.' *Bull. South. Calif. Acad. Sci. 75*, 127-31

Ferguson, A. (1980). *Biochemical Systematics and Evolution.* Blackie, Glasgow

Feuth-de Bruijn, E. and Sevenster, P. (1983). 'Parental Reactions to Young in Sticklebacks (*Gasterosteus aculeatus* L.). *Behaviour 83*, 186-203

Fish, M.P. (1954). 'The Character and Significance of Sound Production among Fishes of the Western North Atlantic.' *Bull. Bingham Oceanogr. Coll. 14*, 1-109

FitzGerald, G.J. (1983). 'The Reproductive Ecology and Behaviour of Three Sympatric Sticklebacks (Gasterosteidae) in a Salt Marsh.' *Biol. Behav. 8*, 67-79

FitzGerald, G.J. and Dutil, J.-D. (1981). 'Evidence for Differential Predation on an Estuarine Stickleback Community.' *Can. J. Zool. 59*, 2394-95

Foster, J.R. (1977). 'The Role of Breeding Behaviour and Habitat Preferences in the Reproductive Isolation of Three Allopatric Populations of Ninespine Stickleback, *Pungitius pungitius.' Can. J. Zool. 55*, 1601-11

Fry, F.E.J. (1971). 'The Effect of Environmental Factors on the Physiology of Fish.' In *Fish Physiology*, Vol. VI (W.S. Hoar and D.J. Randall, eds) pp. 1-98. Academic Press, London

Fryer, G. and Iles, T.D. (1972). *The Cichlid Fishes of the Great Lakes of Africa.* Oliver and Boyd, Edinburgh

Gadgil, M. and Bossert, W.H. (1970). 'Life Historical Consequences of Natural Selection.' *Amer. Nat. 104*, 1-24

Garside, E.T., Heinze, D.G. and Barbour, S.E. (1977). 'Thermal Preference in Relation to Salinity in the Threespine Stickleback, *Gasterosteus aculeatus* L., with an Interpretation of its Significance.' *Can. J. Zool. 53*, 590-4

Gee, J.H., Tallman, R.F. and Smart, H.J. (1978). 'Reactions of some Great Plains Fishes to Progressive Hypoxia.' *Can. J. Zool. 56*, 1962-6

Gerking, S.D. (1980). 'Fish Reproduction and Stress.' In *Environmental Physiology of Fishes* (M.A. Ali, ed.) pp. 569-87. Plenum, New York

Gibson, R.M. (1980). 'Optimal Prey-size Selection by Three-spined

Sticklebacks (*Gasterosteus aculeatus*): a Test of the Apparent Size Hypothesis.' *Z. Tierpsychol. 52*, 291-307

Gilbert, F.F. and Nancekivell, E.G. (1982). 'Food Habits of Mink (*Mustela vison*) and Otter (*Lutra canadensis*) in Northeastern Alberta.' *Can. J. Zool. 60*, 1275-81

Giles, N. (1983a). 'The Possible Role of Environmental Calcium Levels during the Evolution of Phenotypic Diversity in Outer Hebridean Populations of the Three-spined Stickleback, *Gasterosteus aculeatus.' J. Zool., Lond. 199*, 535-44

Giles, N. (1983b). 'Behavioural Effects of the Parasite *Schistocephalus solidus* (Cestoda) on an Intermediate Host, the Three-spined Stickleback, *Gasterosteus aculeatus.' Anim. Behav. 31*, 1192-4

Giles, N. (1984). 'Implications of Parental Care of Offspring for the Anti-predator Behaviour of Adult Male and Female Three-spined Sticklebacks, *Gasterosteus aculeatus* L.' In *Fish Reproduction: Strategies and Tactics* (G.W. Potts and R.J. Wootton, eds) pp. 275-289. Academic Press, London

Golubev, A.V. and Marusov, E.A. (1979). ['Studies on the Chemical Reception of Three-spined Sticklebacks (*Gasterosteus aculeatus* L.) of the White Sea during the Reproductive Period.'] *Biol. Nauki 9*, 33-35. (In Russian.)

Gordon, D.M. and Rau, M.E. (1982). 'Possible Evidence for Mortality Induced by the Parasite *Apatemon gracilis* in a Population of Brook Sticklebacks (*Culaea inconstans*).' *Parasitology 84*, 41-7

Goyens, J. and Sevenster, P. (1976). 'Influence du Facteur Hereditaire et de la Densité de Population sur l'Ontogenèse de l'Aggressivité chez l'Epinoche (*Gasterosteus aculeatus* L.).' *Neth. J. Zool. 26*, 427-31

Greenbank, J. and Nelson, P. (1959). 'Life History of the Three-spined Stickleback *Gasterosteus aculeatus* Linnaeus in Karluk Lake and Bare Lake, Kodiak Island, Alaska.' *U.S. Fish Wildlife Serv. Bull. 153*, 537-59

Griswold, B.L. and Smith, L.L. jr (1972). 'Early Survival and Growth of the Nine-spined Stickleback, *Pungitius pungitius.' Trans. Am. Fish. Soc. 101*, 350-2

Griswold, B.L. and Smith, L.L. jr (1973). 'The Life History and Trophic Relationship of the Ninespine Stickleback, *Pungitius pungitius*, in the Apostle Islands area of Lake Superior.' *Fish. Bull. 71*, 1039-60

Gross, H.P. (1977). 'Adaptive Trends of Environmentally Sensitive Traits in 3-spined Stickleback, *Gasterosteus aculeatus* L.' *Z. zool. Syst. Evol. 15*, 252-77

Gross, H.P. (1978). 'Natural Selection by Predators on the Defensive Apparatus of the Three-spined Stickleback, *Gasterosteus aculeatus* L.' *Can. J. Zool. 56*, 398-413

Gross, H.P. (1979). 'Geographic Variation in European Ninespine Sticklebacks, *Pungitius pungitius.' Copeia* 1979, 405-12

Gross, M.R. (1984). 'Sunfish, Salmon, and the Evolution of Alternative Reproductive Strategies and Tactics in Fishes.' In *Fish Reproduction:*

Strategies and Tactics (G.W. Potts and R.J. Wootton, eds) pp. 55-75. Academic Press, London

Gutz, M. (1970). 'Experimentelle Untersuchungen zur Salzadaptation Verschiedener Rassen des Dreistachligen Stichlings (*Gasterosteus aculeatus* L.).' *Int. Rev. ges. Hydrobiol. 55*, 845-94

Hagen, D.W. (1967). 'Isolating Mechanisms in Three-spine Sticklebacks (*Gasterosteus*).' *J. Fish. Res. Bd Can. 24*, 1637-92

Hagen, D.W. (1973). 'Inheritance of Numbers of Lateral Plates and Gill Rakers in *Gasterosteus aculeatus*.' *Hereditary 30*, 303-12

Hagen, D.W. and Blouw, D.M. (1983). 'Heritability of Dorsal Spines in the Fourspine Stickleback (*Apeltes quadracus*).' *Hereditary 50*, 275-81

Hagen, D.W. and Gilbertson, L.G. (1972). 'Geographical Variation and Environmental Selection in *Gasterosteus aculeatus* L. in the Pacific Northwest, America.' *Evolution 26*, 32-51

Hagen, D.W. and Gilbertson, L.G. (1973a). 'Selective Predation and the Intensity of Selection Acting upon the Lateral Plates of Three-spine Sticklebacks.' *Hereditary 31*, 75-84

Hagen, D.W. and Gilbertson, L.G. (1973b). 'The Genetics of the Plate Morphs in Freshwater Three-spine Sticklebacks.' *Hereditary 31*, 75-84

Hagen, D.W. and McPhail, J.D. (1970). 'The Species Problem within *Gasterosteus aculeatus* on the Pacific Coast of North America.' *J. Fish. Res. Bd Can. 27*, 147-55

Hagen, D.W. and Moodie, G.E.E. (1979). 'Polymorphism for Breeding Colours in *Gasterosteus aculeatus* I. Their Genetics and Geographical Distribution.' *Evolution 33*, 641-8

Hagen, D.W. and Moodie, G.E.E. (1982). 'Polymorphism for Plate Morphs in *Gasterosteus aculeatus* on the East Coast of Canada and an Hypothesis for their Global Distribution.' *Can. J. Zool. 60*, 1032-42

Hagen, D.W., Moodie, G.E.E. and Moodie, P.F. (1980). 'Polymorphism for Breeding Colours in *Gasterosteus aculeatus* II. Reproductive Success as a Result of Convergence for Threat Display.' *Evolution 34*, 1050-9

Hale, P.A. (1965). 'The Morphology and Histology of the Digestive Systems of Two Freshwater Teleosts, *Poecilia reticulata* and *Gasterosteus aculeatus*.' *J. Zool., Lond. 146*, 132-49

Hay, D.E. (1974). 'Ecological Genetics of Threespine Sticklebacks (*Gasterosteus*).' PhD Thesis, University of British Columbia

Hay, D.E. and McPhail, J.D. (1975). 'Mate Selection in Three-spine Sticklebacks (*Gasterosteus*).' *Can.J. Zool. 53*, 441-50

Heller, R. and Milinski, M. (1979). 'Optimal Foraging of Sticklebacks on Swarming Prey.' *Anim. Behav. 27*, 1127-41

Hennig, R. and Zander, C.D. (1981). 'Zur Biologie und Nahrung von Keinfischen des Nord-und Ostsee-Bereichs. 3. Die Besiedlung eines Suesswasserwatts der Elbe durch Euryhaline Fische.' *Arch. Hydrobiol. 43*, 487-505

Heuts, M.J. (1945). 'La Regulation Minerale en Fonction de la

Temperature chez *Gasterosteus aculeatus*. Son Importance au Point de Vue de la Zoogeographie de l'Espèce.' *Ann. Soc. Roy. Zool. Belg. 76*, 88-99

Heuts, M.J. (1947). 'Experimental Studies on Adaptive Evolution in *Gasterosteus aculeatus* L.' *Evolution* 1, 89-102

Heuts, M.J. (1956). 'Temperature Adaption in *Gasterosteus aculeatus* L.' *Publ. Staz. Zool. Napoli 28*, 44-61

Hirshfield, M.F. (1980). 'An Experimental Analysis of Reproductive Effort and Cost in the Japanese Medaka, *Oryzias latipes*.' *Ecology 61*, 282-92

Hochachka, P. and Somero, G.N. (1973). *Strategies of Biochemical Adaptation*. Saunders, London

Honma, Y., Shioda, S. and Yoshie, S. (1977). 'Changes in the Thyroid Gland Associated with the Diadromous Migration of the Threespine Stickleback *Gasterosteus aculeatus*.' *Jap. J. Ichthyol. 24*, 17-25

Hoogland, R.D. (1951). 'On the Fixing-Mechanism in the Spines of *Gasterosteus aculeatus* L.' *Konink. Neder. Acad. weten. Proc. ser. C. 54*, 171-80

Hoogland, R.D., Morris, D. and Tinbergen, N. (1957). 'The Spines of Sticklebacks (*Gasterosteus* and *Pygosteus*) as a Means of Defence against Predators (*Perca* and *Esox*).' *Behaviour 10*, 205-37

Hopkins, C.A. (1959). 'Seasonal Variation in the Incidence and Development of the Cestode *Proteocephalus filicollis* (Rud, 1810) in *Gasterosteus aculeatus* (L. 1776).' *Parasitology 49*, 529-42

Horn, H.S. (1978). 'Optimal Tactics of Reproduction and Life-history.' In *Behavioural Ecology: An Evolutionary Approach* (J.R. Krebs and N.B. Davies, eds) pp. 411-429. Blackwell, Oxford

Hubbell, S.P. (1971). 'Of Sowbugs and Systems: the Ecological Bioenergetics of a Terrestrial Isopod.' In *Systems Analysis and Simulation in Ecology*, Vol. I (B.C. Patten, ed.) pp. 269-324. Academic Press, London

Huntingford, F.A. (1976a). 'The Relationship between Anti-predator Behaviour and Aggression among Conspecifics in the Three-spined Stickleback, *Gasterosteus aculeatus*.' *Anim. Behav. 24*, 245-60

Huntingford, F.A. (1976b). 'A Comparison of the Reaction of Sticklebacks in Different Reproductive Conditions Towards Conspecifics and Predators.' *Anim. Behav. 24*, 694-7

Huntingford, F.A. (1976c). 'An Investigation of the Territorial Behaviour of the Three-spined Stickleback (*Gasterosteus aculeatus*) Using Principal Components Analysis.' *Anim. Behav. 24*, 822-34

Huntingford, F.A. (1977). 'Inter- and Intraspecific Aggression in Male Sticklebacks.' *Copeia* 1977, 158-9

Huntingford, F.A. (1979). 'Pre-breeding Aggression in Male and Female Three-spined Sticklebacks (*Gasterosteus aculeatus*).' *Aggress. Behav. 5*, 51-8

Huntingford, F.A. (1981). 'Further Evidence for an Association between Lateral Scute Number and Aggressiveness in the Threespine Stickleback, *Gasterosteus aculeatus*.' *Copeia* 1981, 717-19

Huntingford, F.A. (1982). 'Do Inter- and Intraspecific Aggression Vary in Relation to Predation Pressure in Sticklebacks?' *Anim. Behav. 30*, 909-916

Hynes, H.B.N. (1950). 'The Food of Freshwater Sticklebacks (*Gasterosteus aculeatus* and *Pygosteus pungitius*) with a Review of the Methods Used in Studies of the Food of Fish.' *J. Anim. Ecol. 19*, 36-58

Iersel, J.J.A. van (1953). 'An Analysis of Parental Behaviour of the Male Three-spined Stickleback (*Gasterosteus aculeatus* L.).' *Behaviour Suppl. 3*, 1-159

Iersel, J.J.A. van (1958). 'Some Aspects of Territorial Behaviour of the Male Three-spined Stickleback.' *Arch. neerl. Zool. 13*, Suppl. 1, 381-400

Jobling, M. (1981). 'The Influence of Feeding on the Metabolic Rate of Fishes: a Short Review.' *J. Fish Biol. 18*, 385-400

Jobling, M. (1983a). 'Growth Studies with Fish – Overcoming the Problems of Size Variation.' *J. Fish Biol. 22*, 153-7

Jobling, M. (1983b). 'Towards an Explanation of Specific Dynamic Action (SDA).' *J. Fish Biol. 23*, 549-55

Jones, D.H. and John, A.G. (1978). 'A Three-spined Stickleback, *Gasterosteus aculeatus* L. from the North Atlantic.' *J. Fish Biol. 13*, 231-6

Jones, J.R.E. (1964). *Fish and River Pollution*. Butterworths, London

Jones, J.W. and Hynes, H.B.N. (1950). 'The Age and Growth of *Gasterosteus aculeatus*, *Pygosteus pungitius* and *Spinachia vulgaris*, as Shown by their Otoliths.' *J. Anim. Ecol. 19*, 59-73

Jordan, C.M. and Garside, E.T. (1972). 'Upper Lethal Temperatures of Threespine Stickleback, *Gasterosteus aculeatus* L. in Relation to Thermal and Osmotic Acclimation, Ambient Salinity and Size.' *Can. J. Zool. 50*, 1405-11

Keenleyside, M.H.A. (1955). 'Some Aspects of Schooling Behaviour of Fish.' *Behaviour 8*, 183-248

Keenleyside, M.H.A. (1979). *Diversity and Adaptation in Fish Behaviour*. Springer-Verlag, Berlin

Ketele, A. and Verheyen, R. (1982). 'Intra- and Interspecific Competition for Space between *Gasterosteus aculeatus* and *Pungitius pungitius*.' *Abst. IV European Ichthyological Congress*, No. 142

Kislalioglu, M. and Gibson, R.N. (1975). 'Field and Laboratory Observations on Prey-size Selection in *Spinachia spinachia* (L.).' *Proc. 9th Europ. Mar. Biol. Symp.* (H. Barnes, ed.) pp. 29-41. Aberdeen University Press, Aberdeen

Kislalioglu, M. and Gibson, R.N. (1976a). 'Prey "Handling Time" and its Importance in Food Selection by the 15-spined Stickleback, *Spinachia spinachia* (L.).' *J. exp. Mar. Biol. Ecol. 25*, 151-8

Kislalioglu, M. and Gibson, R.N. (1976b). 'Some Factors Governing Prey Selection by the 15-spined Stickleback, *Spinachia spinachia* (L.).' *J. exp. Mar. Biol. Ecol. 25*, 159-70

Klinger, S.A., Magnuson, J.J. and Gallepp, G.W. (1982). 'Survival Mechanisms of the Central Mudminnow (*Umbra limi*), Fathead Minnow (*Pimephales promelas*) and Brook Stickleback (*Culaea*

Inconstans) for Low Oxygen in Winter.' *Env. Biol. Fish. 7*, 113-20

Koch, H.J.A. and Heuts, M.J. (1942). 'Influence de l'Hormone Thyroid-ienne sur la Regulation Osmotique chez *Gasterosteus aculeatus* L. forme *gymnurus*.' *Ann. Soc. Roy. Zool. Belg. 73*, 165-72

Krebs, J.R. (1978). 'Optimal Foraging: Decision Rules for Predators.' In *Behavioural Ecology: An Evolutionary Approach* (J.R. Krebs and N.B. Davies, eds) pp. 23-63. Blackwell, Oxford

Krogius, F.V. (1973). 'Population Dynamics and Growth of Young Sockeye Salmon in Lake Dalnee.' *Hydrobiologia 43*, 45-51

Krogius, F.V., Krokhin, E.M. and Menshutkin, V.V. (1972). 'The Modelling of the Ecosystem of Lake Dalnee on an Electronic Computer.' In *Productivity Problems of Freshwaters* (Z. Kajak and A. Hillbricht-Ilkowska, eds) pp. 149-164. PWN, Warsaw and Krakow.

Krokhin, E.M. (1957). 'Determination of the Diurnal Food Rations of Young Sockeye and Three-spined Stickleback Using the Respiration Method.' *Izv. TINRO 44*, 97-110

Krokhin, E.M. (1970). 'Estimation of the Biomass and Abundance of the Threespine Stickleback (*Gasterosteus aculeatus*) in Lake Dal'neye Based on the Food Consumption of Plankton-feeding Fishes.' *J. Ichthyol. 10*, 471-5

Kronnie, G.Te., Tatarczuch, L., Raamsdonk, W. van and Kilarski, W. (1983). 'Muscle Fibre Types in the Myotome of Stickleback *Gasterosteus aculeatus* L.: a Histochemical, Immunohistochemical and Ultrastructural Study.' *J. Fish Biol. 22*, 303-16

Kynard, B.E. (1978a). 'Breeding Behavior of a Lacustrine Population of Threespine Sticklebacks (*Gasterosteus aculeatus* L.).' *Behaviour 67*, 178-207

Kynard, B.E. (1978b). 'Nest Desertion of Male *Gasterosteu aculeatus*.' *Copeia* 1978, 702-3

Kynard, B.E. (1979a). 'Nest Habitat Preference of Low Plate Number Morphs in Threespine Sticklebacks (*Gasterosteus aculeatus*).' *Copeia* 1979, 525-8

Kynard, B.E. (1979b). 'Population Decline and Change in Frequencies of the Lateral Plates in Threespine Sticklebacks (*Gasterosteus aculeatus*).' *Copeia* 1979, 635-8

Kynard, B. and Curry, K. (1976). 'Meristic Variation in Threespine Stickleback, *Gasterosteus aculeatus*, from Auke Lake, Alaska.' *Copeia* 1976, 811-13

Lam, T.J. (1972). 'Prolactin and Hydromineral Regulation in Fishes.' *Gen. Comp. Endocrin. Suppl. 3*, 328-38

Lam, T.J., Nagahama, Y., Chan, K. and Hoar, W.S. (1978). 'Overripe Eggs and Postovulatory Corpora Lutea in the Threespine Stickleback, *Gasterosteus aculeatus* L., Form *Trachurus*.' *Can. J. Zool. 56*, 2029-36

Lam, T.J., Chan, K. and Hoar, W.S. (1979). 'Effect of Progesterone and Estradiol-17B on Ovarian Fluid Secretion in the Threespine Stickleback, *Gasterosteus aculeatus* L. Form *Trachurus*.' *Can. J. Zool. 57*, 468-71

Larson, G.L. (1976). 'Social Behavior and Feeding Ability of Two

Phenotypes of *Gasterosteus aculeatus* in Relation to Their Spatial and Trophic Segregation in a Temperate Lake.' *Can. J. Zool. 54*, 107-21

Leatherland, J.F. (1970). 'Seasonal Variation in the Structure and Ultrastructure of the Pituitary in the Marine Form (*Trachurus*) of the Threespine Stickleback, *Gasterosteus aculeatus* L.' *Z. Zellforsch. 104*, 301-36

Le Cren, E.D. (1951). 'The Length-Weight Relationship and Seasonal Cycle in Gonad Weight and Condition in the Perch (*Perca fluviatilis*).' *J. Anim. Ecol. 20*, 201-19

Lemmetyinen, R. (1973). 'Feeding Ecology of *Sterna paradisea* Pontopp. and *S. hirundo* L. in the Archipelago of Southwest Finland.' *Ann. Zool. Fenn. 10*, 507-25

Lemmetyinen, R. and Mankki, J. (1975). 'The Three-spined Stickleback (*Gasterosteus aculeatus*) in the Food Chains of the Northern Baltic.' *Proc. 3rd Baltic Symp. Mar. Biol. Merentutkimuslait. Julk. 239*, 155-61

Lester, R.J.G. (1971). 'The Influence of *Schistocephalus* Plerocercoids on the Respiration of *Gasterosteus* and a Possible Resulting Effect on the Behaviour of the Fish.' *Can. J. Zool. 49*, 361-6

Lewis, D.B., Walkey, M. and Dartnall, H.J.G. (1972). 'Some Effects of Low Oxygen Tensions on the Distribution of the Three-spined Stickleback *Gasterosteus aculeatus* L. and the Nine-spined Stickleback *Pungitius pungitius* (L.).' *J. Fish Biol. 4*, 103-8

Lewontin, R.C. (1974). *The Genetic Basis of Evolutionary Change.* Columbia University Press, New York

Li, S.K. and Owings, D.H. (1978a). 'Sexual Selection in the Three-spined Stickleback: I Normative Observations.' *Z. Tierpsychol. 46*, 359-71

Li, S.K. and Owings, D.H. (1978b). 'Sexual Selection in the Three-spined Stickleback: II Nest Raiding during the Courtship Phase.' *Behaviour 64*, 298-304

Lindsey, C.C. (1962). 'Experimental Study of Meristic Variation in a Population of Threespine Sticklebacks, *Gasterosteus aculeatus*.' *Can. J. Zool. 40*, 271-312

Lindsey, C.C. (1966). 'Body Size of Poikilotherm Vertebrates at Different Latitudes.' *Evolution 20*, 456-65

Lindsey, C.C. (1978). 'Form, Function and Locomotory Habits in Fish.' In *Fish Physiology*, Vol. VII (W.S. Hoar and D.J. Randall, eds) pp. 1-100. Academic Press, London

Loew, E.R. and Lythgoe, J.N. (1978). 'The Ecology of Cone Pigments in Teleost Fishes.' *Vision Res. 18*, 715-22

Love, R.M. (1980). *The Chemical Biology of Fishes, Vol. 2.* Academic Press, London

Lythgoe, J.N. (1979). *The Ecology of Vision.* Clarendon Press, Oxford

McKenzie, J.A. and Keenleyside, M.H.A. (1970). 'Reproductive Behavior of Ninespine Sticklebacks (*Pungitius pungitius* (L.)) in South Bay, Manitoulin, Ontario.' *Can. J. Zool. 48*, 55-61

Maclean, J. (1980). 'Ecological Genetics of Threespine Sticklebacks in Heisholt Lake.' *Can. J. Zool. 58*, 2026-39

McPhail, J.D. (1969). 'Predation and the Evolution of a Stickleback (*Gasterosteus*).' *J. Fish. Res. Bd Can. 26*, 3183-208

McPhail, J.D. (1977). 'Inherited Interpopulation Differences in Size at First Reproduction in Threespine Stickleback, *Gasterosteus aculeatus*.' *Hereditary 38*, 53-60

McPhail, J.D. and Hay, D.E. (1983). 'Differences in Male Courtship in Freshwater and Marine Sticklebacks (*Gasterosteus aculeatus*).' *Can. J. Zool. 61*, 292-7

McPhail, J.D. and Peacock, S.D. (1983). 'Some Effects of the Cestode (*Schistocephalus solidus*) on Reproduction in the Threespine Stickleback (*Gasterosteus aculeatus*): Evolutionary Aspects of a Host-Parasite Interaction.' *Can. J. Zool. 61*, 901-8

Majkowski, J. and Waiwood, K.G. (1981). 'A Procedure for Evaluating the Food Biomass Consumed by a Fish Population.' *Can. J. Fish. Aquat. Sci. 38*, 1199-208

Mann, R.H.K. (1971). 'The Populations, Growth and Production of Fish in Four Small Streams in Southern England.' *J. Anim. Ecol. 40*, 155-90

Manzer, J.I. (1976). 'Distribution, Food and Feeding of the Three-spine Stickleback, *Gasterosteus aculeatus*, in Great Central Lake, Vancouver Island, with Comments on Competition for Food with Juvenile Sockeye Salmon, *Oncorhynchus nerka*.' *Fish. Bull. 74*, 647-68

Maskell, M., Parkin, D.T. and Verspoor, E. (1977). 'Apostatic Selection by Sticklebacks upon a Dimorphic Prey.' *Hereditary 39*, 83-9

Maynard Smith, J. (1977). 'Parental Investment: a Prospective Analysis.' *Anim. Behav. 25*, 1-9

Maynard Smith, J. (1982). *Evolution and the Theory of Games*. Cambridge University Press, Cambridge

Meakins, R.H. (1974a). 'The Bioenergetics of the *Gasterosteus/Schistocephalus* Host-Parasite System.' *Pol. Arch. Hydrobiol. 21*, 455-66

Meakins, R.H. (1974b). 'A Quantitative Approach to the Effects of the Plerocercoid of *Schistocephalus solidus* Muller 1776 on the Ovarian Maturation of the Three-spined Stickleback *Gasterosteus aculeatus* L.' *Z. Parasitenkd, 44*, 73-9

Meakins, R.H. (1975). 'The Effects of Activity and Season on the Respiration of the Three-spined Stickleback *Gasterosteus aculeatus* L.' *Comp. Biochem Physiol. 51A*, 155-7

Meakins, R.H. (1976). 'Variations in the Energy Content of Freshwater Fish.' *J. Fish Biol. 8*, 221-4

Meakins, R.H. and Walkey, M. (1975). 'The Effects of the Plerocercoid of *Schistocephalus solidus* Muller 1776 (Pseudophyllidea) on the Respiration of the Three-spined Stickleback *Gasterosteus aculeatus* L.' *J. Fish Biol. 7*, 817-24

Meesters, A. (1940). 'Uber die Organization des Gesichsfeldes der Fische.' *Z. Tierpsychol. 4*, 84-149

Milinski, M. (1977a). 'Experiments on the Selection by Predators against Spatial Oddity of their Prey.' *Z. Tierpsychol. 43*, 311-25

Milinski, M. (1977b). 'Do all Members of a Swarm Suffer the Same Predation?' *Z. Tierpsychol. 45*, 373-88

Milinski, M. (1979a). 'Can an Experienced Predator Overcome the Confusion of Swarming Prey more Easily?' *Anim. Behav. 27*, 1122-6

Milinski, M. (1979b). 'An Evolutionarily Stable Feeding Strategy in Sticklebacks.' *Z. Tierpsychol. 51*, 36-40

Milinski, M. (1982). 'Optimal Foraging: the Influence of Intraspecific Competition on Diet Selection.' *Behav. Ecol. Sociobiol. 11*, 109-15

Milinski, M. (1984). 'Competitive Resource Sharing: an Experimental Test of a Learning Rule for ESSs.' *Anim. Behav. 32*, 233-42

Milinski, M. and Heller, R. (1978). 'Influence of a Predator on the Optimal Foraging Behaviour of Sticklebacks (*Gasterosteus aculeatus* L.).' *Nature 275*, 642-4

Milinski, M. and Lowenstein, C. (1980). 'On Predator Selection against Abnormalities of Movement. A Test of an Hypothesis. *Z. Tierpsychol.*' *53*, 325-40

Miller, P.J. (1979a). 'A Concept of Fish Phenology.' *Symp. zool. Soc. Lond. 44*, 1-28

Miller, P.J. (1979b). 'Adaptiveness and Implications of Small Size in Teleosts.' *Symp. zool. Soc. Lond. 44*, 263-306

Miller, R.R. and Hubbs, C.L. (1969). 'Systematics of *Gasterosteus aculeatus* with Particular Reference to Intergradation and Introgression along the Pacific Coast of North America: a Commentary on a Recent Contribution.' *Copeia 1969*, 52-9

Molenda, E. and Fiedler, K. (1971). 'Die Wirkung von Prolaktin auf das Verhalten von Stichlingsmannen (*Gasterosteus aculeatus* L.).' *Z. Tierpsychol. 28*, 463-74

Moodie, G.E.E. (1972a). 'Morphology, Life History and Ecology of an Unusual Stickleback (*Gasterosteus aculeatus*) in the Queen Charlotte Islands, Canada.' *Can. J. Zool. 50*, 721-32

Moodie, G.E.E. (1972b). 'Predation, Natural Selection and Adaptation in an Unusual Threespine Stickleback.' *Hereditary 28*, 155-67

Moodie, G.E.E. (1977). 'Meristic Variation, Asymmetry, and Aspects of the Habitat of *Culaea inconstans* (Kirtland), the Brook Stickleback in Manitoba.' *Can. J. Zool. 55*, 398-404

Moodie, G.E.E. and Reimchen, T.E. (1976). 'Phenetic Variation and Habitat Differences in *Gasterosteus* Populations of the Queen Charlotte Islands.' *Syst. Zool. 25*, 49-61

Moodie, G.E.E., McPhail, J.D. and Hagen, D.W. (1973). 'Experimental Demonstration of Selective Predation on *Gasterosteus aculeatus*.' *Behaviour 47*, 95-105

Moore, J.W. and Moore, I.A. (1976). 'The Basis of Food Selection in Some Estuarine Fishes, Eels, *Anguilla anguilla* (L.)., Whiting, *Merlangius merlangius* (L.), Sprat, *Sprattus sprattus* (L.), and Stickleback, *Gasterosteus aculeatus* L.' *J. Fish Biol. 9*, 375-90

Morris, D. (1958). 'The Reproductive Behaviour of the Ten-spined Stickleback (*Pygosteus pungitius* L.).' *Behaviour Suppl. 6*, 1-154

Mourier, J.-P. (1970). 'Structure Fine du Rein de l'Epinoche (*Gasterosteus aculeatus* L.) au Cours de sa Transformation Muqueuse.'

Z. Zellforsch. 106, 232-50

Muckensturm, B. (1967). 'L'Epinoche et les Leurres, Complexité de la Reaction.' *C.R. Acad. Sci. Paris (D) 264*, 745-748

Muckensturm, B. (1968). 'La Reaction des Epinoches aux Leurres ne Provient pas d'une Confusion avec un Congère.' *C.R. Acad. Sci. Paris (D) 266*, 2114-16

Muckensturm, B. (1978). 'Reactions to Models in Three-spined Sticklebacks Living in Groups.' *Biol. Behav. 3*, 233-42

Mullem, P.J. van (1967). 'On Synchronisation in the Reproduction of the Stickleback (*Gasterosteus aculeatus* L.' Forma *Leiura* Cuv.). *Arch. neerl. Zool. 17*, 258-74

Mullem, P.J. van and Vlugt, J. van der (1964). 'On the Age, Growth and Migration of the Anadromous Stickleback *Gasterosteus aculeatus* L. Investigated in Mixed Populations.' *Arch neerl. Zool. 16*, 111-39

Munzing, J. (1959). 'Biologie, Variabilität und Genetik von *Gasterosteus aculeatus* L. (Pisces). Untersuchungen im Elbegebiet.' *Int. Rev. Ges. Hydrobiol. 44*, 317-82

Munzing, J. (1963). 'The Evolution of Variation and Distributional Patterns in European Populations of the Three-spined Stickleback, *Gasterosteus aculeatus*.' *Evolution 17*, 320-32

Munzing, J. (1972). 'Polymorphe Populationen von *Gasterosteus aculeatus* L. (Pisces, Gasterosteidae) in Sekundaren Intergradioszonen der Deutschen Bucht und Benachbarter Gebiete.' *Faun. Okol. Mitt. 4*, 69-84

Nelson, J.S. (1968). 'Salinity Tolerance of Brook Sticklebacks, *Culaea inconstans*, Freshwater Ninespine Sticklebacks, *Pungitius pungitius*, and Freshwater Fourspine Sticklebacks, *Apeltes quadracus*.' *Can. J. Zool. 46*, 663-7

Nelson, J.S. (1977). 'Evidence for a Genetic Basis for Absence of a Pelvic Skeleton in Brook Stickleback, *Culaea inconstans*, and Notes on the Geographical Distribution and Origin of Loss.' *J. Fish. Res. Bd Can. 34*, 1314-20

Nelson, K. (1965). 'After Effects of Courtship in the Male Three-spined Stickleback.' *Z. vergl. Physiol. 50*, 569-97

O'Brien, W.J., Slade, N.A. and Vinyard, G.L. (1976). 'Apparent Size as the Determinant of Prey Selection by Bluegill Sunfish (*Lepomis macrochirus*).' *Ecology 57*, 1304-30

Ohguchi, O. (1978). 'Experiments on the Selection against Colour Oddity of Water Fleas by Three-spine Sticklebacks.' *Z. Tierpsychol. 47*, 254-67

Ohguchi, O. (1981). 'Prey Density and Selection against Oddity by Three-spined Sticklebacks.' *Adv. Ethol. 23*, 1-79

Ollevier, F. and Covens, M. (1982). 'Oestradiol-induced Female Specific Proteins in the Plasma and in the Oocytes of *Gasterosteus aculeatus* f. *trachurus*. An Immunological Approach.' In *Reproductive Physiology of Fish* (C.J.J. Richter and H.J.Th. Goos, eds) pp. 159-60. Pudoc, Wageningen

Orr, T.S.C., Hopkins, C.A. and Charles, G.H. (1969). 'Host Specificity and Rejection of *Schistocephalus solidus*.' *Parasitology 59*, 683-90

Paepke, H.-J. (1970). 'Studien zur Okologie, Variabilität und Populationsstruktur des Dreistachligen und Neunstachligen Stichlings.' *Veroff. Bez-Muz., Potsdam 21*, 5-48

Paepke, H.-J. (1982). 'Phanogeographische Struckturen in den Susswasserpopulationen von *Gasterosteus aculeatus* L. (Pisces, Gasterosteidae) in der DDR und ihre Evolutionsbiologischen Aspekte.' *Mitt. zool. Mus. Berlin 58*, 269-328

Parin, N.V. (1968). *Ichthyofauna of the Epipelagic Zone*. Israel Prog. Sci. Tranls., 1970, Jerusalem

Parks, J.R. (1982). *A Theory of Feeding and Growth of Animals*. Springer-Verlag, Berlin

Pascoe, D. and Cram, P. (1977). 'The Effect of Parasitism on the Toxicity of Cadmium to the Three-spined Stickleback, *Gasterosteus aculeatus* L.' *J. Fish Biol. 10*, 467-72

Pascoe, D. and Mattey, D. (1977). 'Dietary Stress in Parasitized and Non-parasitized *Gasterosteus aculeatus* L.' *Z. Parasitenkd. 51*, 179-86

Pascoe, D. and Woodworth, J. (1980). 'The Effects of Joint Stress on Sticklebacks.' *Z. Parasitenkd. 62*, 159-63

Peeke, H.V.S. and Veno, A. (1973). 'Stimulus Specificity of Habituated Aggression in the Stickleback (*Gasterosteus aculeatus*).' *Behav. Biol. 8*, 427-32

Peeke, H.V.S. and Veno, A. (1976). 'Response Independent Habituation of Territorial Aggression in the Three-spined Stickleback (*Gasterosteus aculeatus*).' *Z. Tierpsychol. 40*, 53-8

Peeke, H.V.S., Wyers, E.J. and Herz, M.J. (1969). 'Waning of the Aggressive Response to Male Models in the Three-spined Stickleback (*Gasterosteus aculeatus* L.).' *Anim. Behav. 17*, 224-8

Peeke, H.V.S., Figler, M.H. and Blankenship, N. (1979). 'Retention and Recovery of Habituated Territorial Aggressive Behaviour in the Three-spined Stickleback (*Gasterosteus aculeatus* L.): the Role of Time and Nest Reconstruction.' *Behaviour 69*, 171-82

Penczak, T. (1981). 'Ecological Fish Production in Two Small Lowland Rivers in Poland.' *Oecologia 48*, 107-11

Penczak, T. and O'Hara, K. (1983). 'Catch-Effort Efficiency Using Three Small Seine Nets.' *Fish. Mgmt 14*, 83-92

Penczak, T., Suszycka, E. and Molinski, M. (1982). 'Production, Consumption and Energy Transformation by Fish Populations in a Small Lowland River.' *Ekol. pol. 30*, 111-37

Pennycuik, L. (1971a). 'Quantitative Effects of Three Species of Parasites on a Population of Three-spined Sticklebacks, *Gasterosteus aculeatus* L.' *J. Zool., Lond. 165*, 143-62

Pennycuik, L. (1971b). 'Seasonal Variations in the Parasite Infections in a Population of Threespined Sticklebacks, *Gasterosteus aculeatus* L.' *Parasitology 63*, 373-88

Pennycuik, L. (1971c). 'Frequency Distributions of Parasites in a Population of Three-spined Stickleback, *Gasterosteus aculeatus* L., with Particular Reference to the Negative Binomial Distribution.' *Parasitology 63*, 389-406

Pennycuik, L. (1971d). 'Difference in the Parasite Infections in Three-spined Sticklebacks (*Gasterosteus aculeatus* L.) of Different Sex, Age and Size.' *Parasitology 63*, 407-18

Perlmutter, A. (1963). 'Observations on Fishes of the Genus *Gasterosteus* in the Waters of Long Island, New York.' *Copeia* 1963, 168-73

Pianka, E.R. (1976). 'Natural Selection of Optimal Reproductive Tactics.' *Amer. Zool. 16*, 775-84

Pianka, E.R. (1981). 'Competition and Niche Theory.' In *Theoretical Ecology* (R.M. May ed.) pp. 167-96. Blackwell, Oxford

Pitcher, T.J. and Hart, P.J.B. (1982). *Fisheries Ecology*. Croom Helm, London

Popham, E.J. (1966). 'An Ecological Study on the Predatory Action of the Three-spined Stickleback (*Gasterosteus aculeatus* L.).' *Arch. Hydrobiol. 62*, 70-81

Potapova, T.L. (1972). 'The Intraspecies Variability of Threespine Stickleback *Gasterosteus aculeatus*.' *Vopr. Ikhtiol. 12*, 25-40

Pressley, P.H. (1981). 'Parental Effort and the Evolution of Nest Guarding Tactics in the Threespine Stickleback *Gasterosteus aculeatus*.' *Evolution 35*, 282-95

Rajasilta, M. (1980). 'Food Consumption of the Three-spined Stickleback (*Gasterosteus aculeatus* L.).' *Ann. Zool. Fenn. 17*, 123-6

Reimchen, T.E. (1980). 'Spine Deficiency and Polymorphism in a Population of *Gasterosteus aculeatus*, an Adaptation to Predators?' *Can. J. Zool. 58*, 1232-44

Reimchen, T.E. (1982). 'Incidence and Intensity of *Cyathocephalus truncatus* and *Schistocephalus solidus* Infection in *Gasterosteus aculeatus*.' *Can. J. Zool. 60*, 1091-5

Reimchen, T.E. (1983). 'Structural Relationships between Spines and Lateral Plates in Threespine Stickleback (*Gasterosteus aculeatus*.).' *Evolution 37*, 931-46

Reisman, H.M. and Cade, T.J. (1967). 'Physiological and Behavioural Aspects of Reproduction in the Brook Stickleback, *Culaea inconstans*.' *Amer. Midl. Nat. 77*, 257-95

Reist, J.D. (1980a). 'Selective Predation upon Pelvic Phenotypes of Brook Stickleback, *Culaea inconstans*, by Northern Pike, *Esox lucius*.' *Can. J. Zool. 58*, 1245-52

Reist, J.D. (1980b). 'Predation upon Pelvic Phenotypes of Brook Stickleback, *Culaea inconstans* by Selected Invertebrates.' *Can. J. Zool. 58*, 1253-8

Reist, J.D. (1983). 'Behavioural Variation in Pelvic Phenotypes of Brook Stickleback *Culaea inconstans*, in Response to Predation by Northern Pike, *Esox lucius*.' *Env. Biol. Fish 8*, 255-67

Ricker, W.E. (1979). 'Growth Rates and Models.' In *Fish Physiology. Vol VIII* (W.S. Hoar, D.J. Randall and J.R. Brett, eds) pp. 677-743. Academic Press, London

Ridley, M. and Rechten, C. (1981). 'Female Sticklebacks Prefer to Spawn with Males whose Nests contain Eggs.' *Behaviour 76*, 152-61

Røed, K.H. (1979). 'The Temperature Preference of the Three-spined Stickleback, *Gasterosteus aculeatus* L. (Pisces), Collected at Different

Seasons.' *Sarsia 64*, 137-41

Rogers, D.E. (1968). 'A Comparison of the Food of Sockeye Salmon Fry and Threespine Sticklebacks in Wood River Lakes.' *Univ. Wash. Publs 3(NS)*, 1-43

Rogers, D.E. (1973). 'Abundance and Size of Juvenile Sockeye Salmon, *Oncorhynchus nerka* and Associated Species in Lake Aleknagik, Alaska, in Relation to their Environment.' *Fish. Bull. 71*, 1061-75

Rohwer, S. (1978). 'Parent Cannibalism of Offspring and Egg Raiding as a Courtship Strategy.' *Amer. Nat. 112*, 429-40

Rowland, W.J. (1974a). 'Reproductive Behavior of the Four-spine Stickleback *Apeltes quadracus*.' *Copeia* 1974, 183-94

Rowland, W.J. (1974b). 'Ground Nest Construction in the Four-spine Stickleback *Apeltes quadracus*.' *Copeia* 1974, 788-9

Rowland, W.J. (1982a). 'The Effects of Male Nuptial Colouration on Stickleback Aggression: a Re-examination.' *Behaviour 80*, 118-26

Rowland, W.J. (1982b). 'Mate Choice by Male Sticklebacks, *Gasterosteus aculeatus*.' *Anim. Behav. 30*, 1093-8

Rowland, W.J. (1983a). 'Interspecific Aggression and Dominance in *Gasterosteus*.' *Env. Biol. Fish 8*, 269-77

Rowland, W.J. (1983b). 'Interspecific Aggression in Sticklebacks — *Gasterosteus aculeatus* Displaces *Apeltes quadracus*.' *Copeia* 1983, 541-4

Ruby, S.M. and McMillan, D.B. (1970). 'Cyclic Changes in the Testis of the Brook Stickleback *Eucalia inconstans* (Kirtland).' *J. Morph. 145*, 295-318

Sargent, R.C. (1982). 'Territory Quality, Male Quality, Courtship Intrusions, and Female Choice in the Threespine Stickleback *Gasterosteus aculeatus*.' *Anim. Behav. 30*, 364-74

Sargent, R.C. and Gebler, J.B. (1980). 'Effects of Nest Site Concealment on Hatching Success, Reproductive Success, and Paternal Behaviour of the Threespine Stickleback, *Gasterosteus aculeatus*.' *Behav. Ecol. Sociobiol. 7*, 137-42

Schaffer, W.M. (1974). 'Selection for Optimal Life Histories: the Effects of Age Structure.' *Ecology 55*, 291-303

Schaffer, W.M. (1979). 'The Theory of Life-history Evolution and its Application to Atlantic Salmon.' *Symp. zool. Soc. Lond. 44*, 307-26

Schaffer, W.M. and Rosenzweig, M.L. (1977). 'Selection for Optimal Life Histories. II: Multiple Equilibria and the Evolution of Alternative Reproductive Strategies.' *Ecology 58*, 60-72

Schutz, E. (1980). 'Die Wirkung von Untergrund und Nestmaterial auf das Nestbauverhalten des Dreistachligen Stichlings (*Gasterosteus aculeatus*).' *Behaviour 72*, 242-317

Scott, D.B.C. (1979). 'Environmental Timing and the Control of Reproduction in Teleost fish.' *Symp. zool. Soc. Lond. 44*, 105-32

Scott, W.B. and Crossman, E.J. (1973). 'Freshwater Fishes of Canada.' *Bull. Fish. Res. Bd Can.* 184

Segaar, J., De Bruin, J.P.C., Merche, A.P. van der and Merche-Jacobi, M.E. (1983). 'Influence of Chemical Receptivity on Reproductive

Behaviour of the Male Three-spined Stickleback (*Gasterosteus aculeatus*).' *Behaviour 86*, 100-66

Semler, D.E. (1971). 'Some Aspects of Adaptation in a Polymorphism for Breeding Colours in the Threespine Stickleback (*Gasterosteus aculeatus*).' *J. Zool., Lond. 165*, 291-302

Sevenster, P. (1961). 'A Causal Study of a Displacement Activity (Fanning in *Gasterosteus aculeatus* L.).' *Behaviour Suppl. 9*, 1-170

Sevenster, P. and Goyens, J. (1975). 'Experience Social et Role du Sexe dans l'Ontogenèse de l'aggressivité chez l'Épinoche (*Gasterosteus aculeatus*)..' *Neth. J. Zool. 25*, 195-205

Sibly, R.M. (1981). 'Strategies of Digestion and Defaecation.' In *Physiological Ecology* (C.R. Townsend and P. Calow, eds) pp. 109-139. Blackwell, Oxford

Slijkhuis, H. (1978). 'Ultrastructural Evidence for Two Types of Gonadotropin Cells in the Pituitary Gland of the Male Three-spined Stickleback, *Gasterosteus aculeatus*.' *Gen. Comp. Endocrin. 36*, 639-41

Slijkhuis, H. (1980). 'Quantitative Electron Microscopical and Physiological Study on the Hormonal Regulation of Reproductive and Parental Behaviour of the Male Threespined Stickleback (*Gasterosteus aculeatus*).' *Neth. J. Zool. 30*, 530-1

Slobodkin, L.B. and Rapoport, A. (1974). 'An Optimal Strategy of Evolution.' *Quart. Rev. Biol. 49*, 181-200

Smith, R.F.J. (1970). 'Effects of Food Availability on Aggression and Nest Building in Brook Stickleback (*Culaea inconstans*).' *J. Fish. Res. Bd Can. 27*, 2350-5

Stanley, B.V. (1983). 'Effect of Food Supply on Reproductive Behaviour of Male *Gasterosteus aculeatus*.' PhD thesis, University of Wales

Stearns, S.C. (1976). 'Life-history Tactics: a Review of the Ideas.' *Quart. Rev. Biol. 51*, 3-47

Stearns, S.C. (1977). 'The Evolution of Life History Traits.' *Ann. Rev. Ecol. System. 8*, 145-71

Symons, P.E.K. (1965). 'Analysis of Spine-raising in the Male Three-spined Stickleback.' *Behaviour 26*, 1-74

Tanaka, S. and Hoshimo, M. (1979). 'Growth and Maturity of Nine-spine Stickleback, *Pungitius sinensis* Guichenot in the Kamokawa, a Streamlet in Toyama Prefecture.' *Bull. Toyama Sci. Mus. 1*, 19-29

t'Hart, M. (1978). 'A Study of a Short Term Behaviour Cycle: Creeping Through in the Three-spined Stickleback (*Gasterosteus aculeatus* L.). Part 2.' *Behaviour 67*, 1-66

Theisen, B. (1982). 'Functional Morphology of the Olfactory Organ in *Spinachia spinachia* (L.)' (Teleostei, Gasterosteidae). *Acta Zool. 63*, 247-54

Thomas, G. (1974). 'The Influence of Encountering a Food Object on Subsequent Searching Behaviour in *Gasterosteus aculeatus* L.' *Anim. Behav. 22*, 941-52

Thomas, G. (1976). 'Gust or Disgust or the Causes of Alliesthesia.

Motivational Changes upon Exposure to Food Stimuli in *Gasterosteus aculeatus* L.' Doctoral Thesis, University of Groningen

Thomas, G. (1977). 'The Influence of Eating and Rejecting Prey Items upon Feeding Behaviour and Food Searching Behaviour in *Gasterosteus aculeatus* L.' *Anim. Behav. 25*, 52-66

Threlfall, W. (1968). 'A Mass Die-off of Three-spined Sticklebacks (*Gasterosteus aculeatus* L.) Caused by Parasites.' *Can. J. Zool. 46*, 210-34

Thresher, R.E. (1978). 'Territoriality and Aggression in the Threespot Damsel Fish (Pisces; Pomacentridae): an Experimental Study of Causation.' *Z. Tierpsychol. 46*, 401-34

Tinbergen, N. (1951). *The Study of Instinct*. Clarendon Press, Oxford

Tugendhat, B. (1960a). 'The Normal Feeding Behaviour of the Three-spined Stickleback (*Gasterosteus aculeatus* L.).' *Behaviour 15*, 284-318

Tugendhat, B. (1960b). 'The Disturbed Feeding Behaviour of the Three-spined Stickleback: I. Electric Shock is Administered in the Food Area.' *Behaviour 16*, 159-87

Tugendhat, B. (1960c). 'Feeding in Conflict Situations and following Thwarting.' *Science 132*, 896-7

Turnpenny, A.W.H. and Williams, R. (1981). 'Factors Affecting the Recovery of Fish Populations in an Industrial River.' *Envir. Pollut. (Ser. A). 26*, 39-58

Ursin, E. (1967). 'A Mathematical Model of Some Aspects of Fish Growth, Respiration and Mortality.' *J. Fish. Res. Bd Can. 13*, 2355-453

Ursin, E. (1979). 'Principles of Growth in Fishes.' *Symp. zool. Soc. Lond. 44*, 63-87

Veen, Th. van, Ekstrom, P. and Borg, B. (1980). 'The Pineal Complex of the Three-spined Stickleback *Gasterosteus aculeatus* L. A Light-, Electron Microscopic and Fluorescence Histochemical Investigation.' *Cell Tiss. Res. 209*, 11-28

Visser, M. (1981). 'Prediction of Switching and Counter Switching Based on Optimal Foraging.' *Z. Tierpsychol. 55*, 129-38

Visser, M. (1982). 'Prey Selection by Three-spined Stickleback (*Gasterosteus aculeatus* L.).' *Oecologia 55*, 395-402

Walkey, M. (1967). 'The Ecology of *Neoechinorhynchus rutili* (Muller).' *J. Parasitol. 53*, 795-801

Walkey, M. and Meakins, R.H. (1970). 'An Attempt to Balance the Energy Budget of a Host-Parasite System.' *J. Fish Biol. 2*, 361-72

Wallace, R.A. and Selman, K. (1979). 'Physiological Aspects of Oogenesis in Two Species of Sticklebacks, *Gasterosteus aculeatus* L. and *Apeltes quadracus* (Mitchill).' *J. Fish Biol. 14*, 551-64

Warren, C.E. and Davis, G.E. (1967). 'Laboratory Studies on the Feeding Bioenergetics and Growth of Fish.' In *The Biological Basis of Freshwater Fish Production* (S.D. Gerking, ed.) pp. 175-214. Blackwell, Oxford

Wendelaar Bonga, S.E. (1973). 'Morphometrical Analysis with the Light and Electron Microscope of the Kidney of Anadromous Three-spined Stickleback *Gasterosteus aculeatus* form *trachurus* from fresh water and from sea water.' *Z. Zellforsch. 137*, 563-88

Wendelaar Bonga, S.E. (1976). 'The Effect of Prolactin on Kidney Structure of the Euryhaline Teleost *Gasterosteus aculeatus* during Adaptation to Fresh Water.' *Cell Tiss. Res. 166*, 319-38

Wendelaar Bonga, S.E. (1978a). 'The Effects of Changes in External Sodium, Calcium, and Magnesium Concentrations on Prolactin Cells, Skin, and Plasma Electrolytes of *Gasterosteus aculeatus*.' *Gen. Comp. Endocrin. 34*, 265-75

Wendelaar Bonga, S.E. (1978b). 'The Role of Environmental Calcium and Magnesium Ions in the Control of Prolactin Secretion in the Teleost *Gasterosteus aculeatus*.' In *Comparative Endocrinology* (P.J. Gaillard and H.H. Boer, eds) pp. 259-62. Elsevier/North-Holland, Amsterdam

Westerfield, F. (1922). 'The Ability of Mudminnows to form Associations with Sounds.' *Comp. Psychol. 2*, 187-90

Wheeler, A. (1969). *The Fishes of the British Isles and Northwest Europe*. Macmillan, London

Wheeler, A. (1979). *The Tidal Thames*. Routledge and Kegan Paul, London

Williams, G.C. (1966). *Adaptation and Natural Selection*. Princeton University Press, Princeton

Wilz, K.J. (1970a). 'Self-regulation of Motivation in the Three-spined Stickleback (*Gasterosteus aculeatus* L.).' *Nature 266*, 465-6

Wilz, K.J. (1970b). 'Causal and Functional Analysis of Dorsal Pricking and Nest Activity in the Courtship of the Three-spined Stickleback *Gasterosteus aculeatus*.' *Anim. Behav. 18*, 115-24

Wilz, K.J. (1970c). 'The Disinhibition Interpretation of the 'Displacement' Activities during Courtship in the Three-spined Stickleback, *Gasterosteus aculeatus*.' *Anim. Behav. 18*, 682-7

Wilz, K.J. (1970d). 'Reproductive Isolation in Two Species of Stickleback (Gasterosteidae).' *Copeia* 1970, 587-90

Wilz, K.J. (1971). 'Comparative Aspects of Courtship Behaviour in the Ten-spined Stickleback, *Pygosteus pungitius* (L.).' *Z. Tierpsychol. 29*, 1-10

Wilz, K.J. (1973). 'Quantiative Difference in the Courtship of Two Populations of Three-spined Sticklebacks, *Gasterosteus aculeatus*.' *Z. Tierpsychol. 33*, 141-6

Wilz, K.J. (1975). 'Cycles of Aggression in Male Three-spined Stickleback (*Gasterosteus aculeatus*).' *Z. Tierpsychol. 39*, 1-7

Winberg, G.C. (1956). 'Rate of Metabolism and Food Requirements of Fishes.' *Fish. Res. Bd Can. Transl. Ser. No. 194*, 1-253

Wootton, R.J. (1971a). 'Measures of the Aggression of Parental Male Three-spined Sticklebacks.' *Behaviour 40*, 228-62

Wootton, R.J. (1971b). 'A Note on the Nest-raiding Behaviour of Male Sticklebacks.' *Can. J. Zool. 49*, 960-2

Wootton, R.J. (1972). 'The Behaviour of the Male Three-spined Stickleback in a Natural Situation: a Quantitative Description.' *Behaviour 41*, 232-41

Wootton, R.J. (1973a). 'The Effect of Size of Food Ration on Egg Production in the Female Three-spined Stickleback, *Gasterosteus aculeatus* L.' *J. Fish Biol. 5*, 89-96

Wootton, R.J. (1973b). 'Fecundity of the Three-spined Stickleback, *Gasterosteus aculeatus* L.' *J. Fish Biol. 5*, 683-8

Wootton, R.J. (1974a). 'The Inter-spawning Interval of the Female Three-spined Stickleback, *Gasterosteus aculeatus*.' *J. Zool., Lond. 172* 331-42

Wootton, R.J. (1974b). 'Changes in the Courtship Behaviour of Female Three-spined Sticklebacks between Spawnings.' *Anim. Behav. 22*, 850-5

Wootton, R.J. (1976). *The Biology of the Sticklebacks*. Academic Press, London

Wootton, R.J. (1977). 'Effect of Food Limitation during the Breeding Season on the Size, Body Components and Egg Production of Female Sticklebacks (*Gasterosteus aculeatus*).' *J. Anim. Ecol. 46*, 823-34

Wootton, R.J. (1979). 'Energy Costs of Egg Production and Environmental Determinants of Fecundity in Teleost Fishes.' *Symp. zool. Soc. Lond. 44*, 133-59

Wootton, R.J. (1984). 'Introduction: Strategies and Tactics in Fish Reproduction.' In *Fish Reproduction: Strategies and Tactics* (G.W. Potts and R.J. Wootton, eds) pp. 1-12. Academic Press, London

Wootton, R.J. and Evans, G.W. (1976). 'Cost of Egg Production in the Three-spined Stickleback (*Gasterosteus aculeatus*).' *J. Fish Biol. 8*, 385-95

Wootton, R.J., Evans, G.W. and Mills, L. (1978). 'Annual Cycle in Female Three-spined Sticklebacks (*Gasterosteus aculeatus* L.) from an Upland and Lowland Population.' *J. Fish Biol. 12*, 331-43

Wootton, R.J., Allen, J.R.M. and Cole, S.J. (1980a). 'Energetics of the Annual Reproductive Cycle in Female Sticklebacks, *Gasterosteus aculeatus* L.' *J. Fish Biol. 17*, 387-94

Wootton, R.J., Allen, J.R.M. and Cole, S.J. (1980b). 'Effect of Body Weight and Temperature on the Maximum Daily Food Consumption of *Gasterosteus aculeatus* L. and *Phoxinus phoxinus* (L.): Selecting an Appropriate Model.' *J. Fish Biol. 17*, 695-705

Worgan, J.P. and FitzGerald, G.J. (1981a). 'Habitat Segregation in a Salt Marsh among Adult Sticklebacks (Gasterosteidae).' *Env. Biol. Fish. 6*, 105-9

Worgan, J.P. and FitzGerald, G.J. (1981b). 'Diel Activity and Diet of Three Sympatric Sticklebacks in Tidal Salt Marsh Pools.' *Can. J. Zool. 59*, 2375-9

Ziuganov, V.V. (1983). 'Genetics of Osteal Plate Polymorphism and Microevolution of Threespine Stickleback (*Gasterosteus aculeatus* L.).' *Theor. Appl. Genet. 65*, 239-46

INDEX